The Newbie Thesis

JON SEARS

ISBN: 147747238X
ISBN-13: 9781477472385

Library of Congress Control Number: 2012908881
CreateSpace, North Charleston, SC

Dedicated to Anne Rice

It was August 11, 2010 when I watched Dan Harris interview the author, Anne Rice. I thought I was prepared. I had my pad and pencil ready to jot down some writing tips but it never happened. Instead, there she was on ABC News© expressing her disappointment with the Catholic and Mormon Churches' for persecuting gays. Maybe it was my receptive state of mind but I experienced a paradigm shift when she mentioned her son was gay. Suddenly, she became a mom pleading for her child. There was a sense of helplessness about her. It seemed as though she needed a unique kind of help; she needed the power of an imaginative pen to open the minds of the persecutors.

Suddenly, I realized she represented millions of moms struggling with some variation of the same challenge. However the other moms did not have the benefit of celebrity, TV cameras and microphones to share their dilemma. This is when I knew I was not going to allow her and all the other moms to fight this battle by themselves. This is when I became an advocate for gay rights.

Her predicament sent me to my digital archives where I retrieved 'The Newbie Thesis' a story that answers the question, 'why are some people gay' in a manner that clearly shows that gays have no say in the matter. If persecutors are angry with someone they should take it up with our lord and savior because he is the one directing the path our species is taking.

The world has never been more ready for a story!

The question, 'Why are some persons Gay?' has nagged our society for centuries as it does today. It's the question families with gay challenges have often asked their physicians and religious leaders and received unconvincing answers. In contrast, *The Newbie Thesis'* answer is so believable it's astonishing! *Click-up* is the sensation people feel when they hear the belt-buckling answer that fits snug like a missing piece to a puzzle.

The story takes place in the year 2055 because the hi-tech laboratory, medical knowledge, and the tools used in the story do not exist today. Readers looking for a story about sex should forget it. *The Newbie Thesis* is about families struggling with tough ambiguous challenges. It's not a gay story nor is it a straight story it's a story that gives hope and dignity to everyone.

The use of accepted scientific theorems gives my theory impeccable credibility making it impossible to disprove without dismissing Darwin's theories as well.

So now those courageous moms have God and Darwin on their side. They also have a powerful argument to speak on their behalf. It's a non-violent voice capable of breeding understanding and opening avenues of thought not yet pursued.

Thank you Ms. Rice.

1. The Inciting Event

Information is power and with it comes the awesome ability to get things done. Proof that humans possess an insatiable thirst for power is a fact, well documented. Tragically, in past years, overzealous scientists have violated the laws of human decency to quench their thirst. Consequently, no one was surprised in 2055 when it happened again in the City of Summervale.

A video phone the size of a ballpoint pen stood on tripod legs and projected the virtual image of a woman to Michael McGowan. Michael had founded the Gay Activists Organization and today he had every reason to be jovial about the way things were going.

"Congratulations, Estelle. Thanks to you, the missing puzzle pieces about the Inquiry are becoming less of a mystery. If our investigation continues at this pace, we're going to force those bastards out of hiding."

"That's good, Michael, but I don't feel comfortable having this photo in my possession. I want you to come and get it...and I mean now."

"That's no problem—" was all Mike had a chance to say before Estelle interrupted.

"Does this mean I'll see you this evening?" she asked while sounding and appearing to be jittery.

"I'll see ya in twenty," said Mike as he swiveled his chair around, expecting to see an office full of busy coworkers. But it was well past closing, and the only person there was his best friend Seth, standing in the doorway.

Seth Bluestein was barely twenty. He was exceptionally bright and he always had whatever book he was reading close, at hand. Footprints from eyeglasses he'd worn since childhood marred the bridge of his nose. He pushed them up and stared at the banner hanging on the wall above Mike's desk. It read "Why Am I Gay?" That question had made Mike an Internet sensation and his quest for an answer was getting plenty of help from his cyber audience.

"Why are you staring at my banner?" "It's on display to remind the staff to stay focused until we find an answer that makes sense, but you know already that."

"Sure I know," said Seth, picking up a woman's picture from Mike's desk to gaze at her. "But I'm curious…does Lovie ever wonder about the answer to our question?"

"My sister is not gay," said Mike, with a tone of finality. He snatched her picture from Seth and returned it to its place on his desk.

"I know she's straight, but she *is* your twin. You shared a womb, but you came out gay. What does that tell us?"

*Whew…*whistled Mike as he exhaled hot air before he spoke. "Lovie and I have discussed this…and we agree that whatever the answer—be it good, bad, or theoretical—gay people and their straight relatives deserve the truth, whatever it is."

"Amen to that and we will find the truth," said Seth. They shook hands and hugged the way ballplayers do before moving on to the next challenge.

1. The Inciting Event

A good poker face—Michael did not have. In fact, good poker players would describe his brow as a "tell." It always wrinkled up something awful whenever Mike tried to figure something out, like now when he asked, "Did you witness my phone call?"

"Some of it," Seth replied.

"That was Estelle. She says she has a photo of several high-ranking members of the Inquiry. Can you believe our luck?"

"Whoa, now that's potent. Does she have the names to match the pictures?" Seth asked.

"We're not that lucky but we're working on it. In the meantime she wants me to pick up the photo right away because she's afraid to e-mail it."

"What about our seven-o'clock appointment? The informant sounds like he has some valuable info," Seth reminded Mike.

"That's no problem. I can swing by Estelle's Shoppe *and* get to the pub in time for our meeting. By chance...did you notice anything unusual about Estelle?"

"You were saying good-bye when I walked up. Why don't you replay her call and I'll watch it with ya?"

Mike stood and started toward the door. "I'd like to...but it's almost six, so I've got to hit the road."

"I think we should stick to our original plan," said Seth.

"Have you *ever* known me to be late? Mike asked. "Tell the truth. Who's always the first on the scene? Hun, you know who to applaud," he boasted as he opened the hallway door.

Suddenly, Mike spun around in the doorway and pointed both index fingers at Seth and went "Pow, pow!" as if firing a pair of toy pistols. As strange as it was, this was Mike's trademark good-bye, a reenactment from childhood. He blew the smoke off his pistols, backed into the hallway, and the door closed behind him.

2. Chatter Software, LLC

It took less than thirty years for Summervale to replace Silicon Valley as the epicenter of modern technology. Technologists started to gather when Chatter, a small nanotech company, introduced the device that revolutionized information technology forever. Known as the PenSet, it became the world's communication standard. One Pen replaced the smart phones, and when coupled with a second Pen, they projected usable virtual images of PCs, laptops, pads, and pods which made the old standard devices obsolete. The pads, especially, with their cumbersome fixed screens, simply could not compete with the adjustable screens offered by the PenSet.

Chatter called the third Pen the ChatBox because it was the universal language translator that tore down the language barriers between nations. Now the people of the world felt as though they all spoke the same language, and because of it, the world moved closer to peace. When not in use, all three Pens clipped onto a user's pocket like ordinary ballpoint pens.

The PenSet transformed Chatter into a world conglomerate. Suppliers and other high-tech companies moved to Summervale in pursuit of business alliances. Job opportunities, marvelous weather, and oceanfront properties attracted well-educated people looking for challenges and rewards. The influx of talent spurred consumer demand, and a bustling economy was born.

Estelle's Curio Shoppe was in an upscale shopping district with art-deco hotels that seemed to radiate warmth. Tall, majestic palm trees lined streets where expensive boutiques displayed stylish apparels in their windows.

The doorbell chimed the moment Mike entered her shop. He was surprised to find the lights dim and no one to greet him. "Hey, Estelle, its Mike…where are you?" he called, as he ventured in farther. He got no answer, so he busied himself with looking for something nice to buy for Lovie. He selected a pink silk scarf and took it to the counter where he waited to purchase it.

Suddenly, he caught the glimpse of a body crumpled on the floor behind the counter. It was Estelle. Instinctively, he kneeled to help her. This was the last thing he remembered before someone knocked him unconscious.

Mike woke up in Estelle's back office sitting in a chair. Stars swirled in his head, and the spot where a batter had used his skull for a baseball ached something awful. He couldn't touch the sore spot; because someone had tied him to the chair. There were ropes around his ankles. His chest was bound to the backrest and his wrists to the arms of the chair.

Angered, Mike struggled to free himself until he heard footsteps coming his way. He looked up to see two huge, gruff-looking thugs with stares that warned him not to say a damn thing. Standing between them was a tall, muscle-bound black Amazon who looked as though she could lift both fellows at the same time. She could easily have been mistaken for a man had not the faint hint of breasts gave her away.

"My name is Konnerman," she said. "I have a question and if you answer it, maybe you can get out of here before my boss arrives."

"Fire away," said Mike.

"Where's the book?"

"Book…what book? Ma'am, I don't have an inkling of what you're talking about…I came here for a picture. I don't know about a book."

"Estelle already surrendered the picture. But where's the book?" Konnerman asked but this time, her impatience was obvious.

"I have no idea, ma'am, but if I did…I swear I'd give it to you," said Mike.

Konnerman yanked him by the collar so hard that she lifted him and the chair off the floor. "Where's the damn book?" she yelled as she vigorously shook him like a rag doll.

Ding-a-ling, the front doorbells chimed. Konnerman set Mike and his chair on the floor just before a tall, dark and sinister figure entered the dimly lit office.

"Michael, this is Dr. Scorn," said Konnerman. "From here on, he will conduct the questioning."

Dr. Liborio Scorn looked to be in his mid-seventies. He had the face, crooked nose, and rounded shoulders of an evil witch. He wore a black derby hat, black leather boots and a black cloak that added a diabolical look to his old wrinkled skin. In a hoarse, wispy voice, he said, "It would be best if you tell me what I want to know before I start my work."

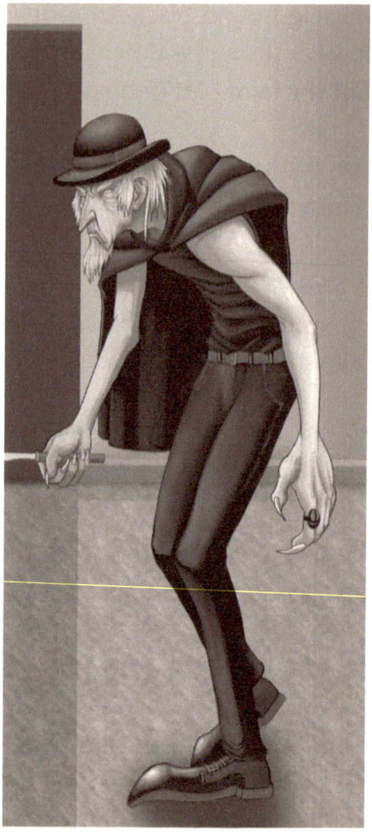

"I've never seen nor heard of your book." Mike pleaded with his words and eyes.

The doctor always carried a lethal stun gun concealed within the oval ring he wore on his finger. He hunched his shoulders up and down as if he was loosening up his muscles. Then he pointed a crooked finger at Mike. The gestures alone were intimidating, but Mike shuddered even more when he recognized the dreaded symbol of the Inquiry on the doctor's arm.

"Aha! I see you've noticed my tattoo," said the doctor. "Good. Then you know I'm from the Inquiry, which means I've had considerable practice questioning stubborn little people like you."

Dr. Scorn, who looked like he hadn't smiled in decades, motioned to one of the thugs, who covered Mike's mouth with several strips of tape. He extracted a black bundle from his cloak and placed it on Estelle's desk. When he unraveled it, Mike could see an assortment of shiny surgical instruments.

The doctor picked up one of the instruments and proudly showed it to Mike. "This is *the* laser-scalpel favored by most surgeons nowadays," he said, fondling and turning it over in his hands as he admired its beauty. Then he pressed the *'on'* button and a laser beam of light in the shape of a scalpel streaked from the handle.

Michael watch in horror as Dr. Scorn adjusted the size of the blade. His throat was so parched he couldn't swallow. Knowing this was his dying moment, he prayed for a swift and painless death.

"I'm told that you have a unique way of saying good-bye," said Dr. Scorn. Then he raised one of Mike's "trigger" fingers and sliced it off before asking his first question.

The tape muted Mike's screams, but tears, saliva, and sweat poured from his body as he pleaded for mercy with his eyes and muffled cries. The angry doctor yelled, "I understand that you want to know why you are gay. Well, I can provide the answer. But first, where…is…my…book?"

Mike's grimaces showed the pain was severe. He couldn't speak, so he grunted and shook his head to mean he was not aware of the book. This prompted Dr. Scorn to cut off the second trigger finger and Michael passed out.

"We're wasting time. He doesn't know anything," said Dr. Scorn, waving for the thugs to take over. Konnerman stepped forward. She stood in front of Mike and carefully put a lock of his curly brown hair back in place, as if she cared. She gripped the top of his head with one hand and his chin with the other. Suddenly, she twisted Mike's head so hard his neck snapped like a dry twig and the interview was over.

3. Summervale Police

A frightened and panic-stricken janitor called 911 at 4:00 a.m. and reported finding two bodies inside Estelle's boutique. Within an hour, the crime scene investigators alerted the Police Chief after identifying Michael as the son of Chatter's founder.

A few minutes later, Lieutenant Muggs' rotary phone rang repeatedly. Eventually, it woke him out of a deep sleep. But even then, Muggs didn't pick up the phone right away. Instead, he watched the damn thing ring several times while he mentally prepared for the inevitable bad news awaiting his attention.

"What is it this time," he growled into the phone.

"...and that's a good morning to you also," said the Chief. "We have a situation that requires your help."

"...as long as it doesn't take more than two weeks. After that...I'm out of here. What' cha got, Chief?"

"...a double murder is a possibility."

Muggs was a grumpy, old-style cop who didn't like bad news or new ideas. But this time he had a legitimate gripe. "Ah Chief, I'm retiring in a few weeks!

For the sake of common decency, give an old-timer a pass on this," he pleaded.

"No can do," said the Chief. "This is a high-profile case and you're our best detective."

"But Chief, the press is going to be on this more than we can imagine," he said.

"That's why I need you. So call whoever you want to continue the case after you retire. I want him or her to start on the case, with you, *this* morning."

What else could Muggs do but accept? He knew that if the murders became an embarrassment, the department would make him the fall guy. *So what else is new?*

"I'll take Officer Perry with me," he said as he crawled out of his warm and comfortable bed.

Five hours later: 9:00 a.m.

Michael's sister, Lovie McGowan, lived in a posh Summervale community called the Ocean Marina. The condominium's architecture had a Spanish influence. The stylish buildings were wrapped in stucco and complemented with red clay tile roofs and relaxing pastel colors.

There were long wharfs with staging berths where ships could dock and load and unload passengers as well as cargo. The piers floated on pontoons that kept them buoyed at the same level as the ships during changing tides.

The police officers parked in the visitor's lot. Muggs flung his heavy laptop, in its carrying case, over his shoulder and they walked the remaining distance to Lovie's Condo.

"This place never fails to impress me. It's always busy," said Muggs, pointing to some eager boaters using a slipway to transfer a trailered boat into the water. "Take a look farther out that way." He pointed. "See…? Look how those harbor taxis are shuttling people between the shores and their yachts my, oh my, what a life."

Muggs shifted his heavy laptop from one shoulder to the other as they walked a path that skirted the edge of manicured yards and hedges.

"Would you like me to carry your laptop the rest of the way, Lieutenant? It looks really heavy," Officer Perry asked.

"Nah, I'm used to it."

"You know, Muggs…you're one of a kind and I'm gonna miss you. Who else would insist on carrying one of those old-fashioned laptops when he could be carrying a PenSet in his pocket?

"This is my baby. She's been with me since day one," said Muggs, patting the carrying-case as if there were a loyal pet inside.

"Yeah, but why put up with all that cumbersome weight?"

"We're here," said Muggs ringing the doorbell and being grateful he didn't have to answer any more questions from the young pup tagging along with him.

"This is a honey of a place. I wonder how much it costs to live here," said Officer Perry.

"I can't count that high—," Muggs was about to explain, when the sound of Lovie's voice came from the backyard.

"We're around here on the patio," she yelled.

The officers followed the sound of her voice. After turning a few corners, they saw two stunning-looking ladies working on a project. The blonde haired woman stepped forward and greeted them. She had the gracefulness

of a predator and a disarming smile. She stood five feet ten and had an athletic build. There was something competitive about her. Maybe it was the way she kept her eyes focused on the police officers when she shook their hands with a firm grip. "I'm Lovie McGowan, and this is my friend and neighbor Raquel Graciano."

"Ladies," the officers replied in unison while touching the rims of their hats. Surveying the surroundings, the officers noticed a Summervale University jacket on top of a stack of textbooks.

"You must be Lieutenant Muggs…and you're Officer Perry, "said Lovie. "Marina Security called and said you were on your way, but they didn't explain the purpose of your visit."

"Didn't your parents call and tell you we were coming?" Lt. Muggs asked.

"I haven't spoken with my parents in seven years. So what would make this day any different?" she asked.

The police officers exchanged quick glances the moment they realized the burden of telling Lovie about Michael's death had fallen on their shoulders.

Lovie and Raquel noticed the officers' reaction and instinctively braced themselves for bad news. "Oh—boy, this is not gonna be good," Raquel whispered.

"Is there a problem?" Lovie asked with a hint of apprehension in her voice.

"Ma'am, would you like to sit down?" Officer Perry asked, gesturing for Lovie to sit in a patio chair, which she did, reluctantly.

"We got a call…early this morning…from a person claiming to have found two bodies in Estelle's boutique," said Lieutenant Muggs.

"Our CSI team was there in no time," said Officer Perry.

"That's fine work, gentlemen, but what does that have to do with me?" Lovie asked nervously.

"Your brother Michael is one of the deceased," said Muggs.

"Oh no…no, this cannot be! I don't believe you!" Lovie exclaimed. She pointed her finger at the police officers and said, "You've made a terrible mistake!"

The officers remained silent and waited for her next reaction. "This is not real. It's a bad dream…it has to be," she said, looking at her neighbor as if Raquel could make the nightmare go away.

"Ma'am, we know this is difficult," said Muggs. "But I want you to know that we scanned the faces of both persons and compared them to our face-identification database."

"Well…?" Lovie asked, still unable to face the unimaginable loss of Michael.

Lt. Muggs explained, "When our computers identified the male as Michael McGowan, we took it a step further before visiting you. We contacted your parents after which your dad met us at the morgue and identified Michael."

"Oh, no, no, no…!" Lovie screamed, leaping from her chair. Suddenly, she grabbed her neck. "I can't breathe," she said. Then her body went limp and she fainted.

The police officers caught her in time to put her back in the chair. Raquel's reactions were equally quick. She ran to get a pitcher of ice water from the fridge. She poured some on a towel and wiped Lovie's face and neck.

When Lovie regained consciousness Raquel gave her a few sips of cold water. After a few minutes, Lovie regained some of her composure.

"Lovie and Michael were very close," said Raquel. I'm going to call her doctor. Is there anything else?"

"Mr. McGowan was unable to answer questions about Lovie's and Mike's most recent activities, so he told us to ask her," said Muggs.

"Ask her what?" Raquel asked.

"Did Mike have any enemies? A list of persons who might have had a score to settle would be helpful," Muggs replied.

"I'll find out for you. But in the meantime, what are *you* doing to find the murderer?" Raquel asked.

"We are reviewing all the computer files of the Gay Activists' staff, as well as Mike's webcasts looking for clues," said Muggs.

"There is one more item," said Officer Perry. "We are looking for Seth Bluestein. We've interviewed all the employees except him…any idea where he might be?"

"I don't know," said Raquel, "but I'll get back with you on that as well. Gentlemen, right now, I could use some help getting Lovie inside the condo."

"Are you going to contact her parents?" Muggs asked Raquel. But Lovie, who had been listening to the conversation, found the strength to answer herself.

"My parents rejected Michael because he was gay. They drove him out of their home and their lives like so many other snobbish parents. Now he's dead. No, Lieutenant…I will not be calling my parents. I will take care of my brother's funeral arrangements myself."

4. Michael's Burial

The McGowans were a devout Catholic family, split by divergent views of homosexuality. Some of their opinions were condemning; others were tolerant. But today, they stood around Michael's gravesite as a family united with one exception—Lovie was a holdout.

Lovie found it a source of strength to take charge of Michael's funeral arrangements. She had planned and paid for everything, but she was not mentally ready to be in the company of those who had brought turmoil and bitter conflicts into his life. She had refused to sit with her family throughout the funeral services. Now, during the burial, she and Raquel stood at the foot of the gravesite while the rest of the family gathered around the other sides.

"I can sense your family's tension, and I've never met them," Raquel whispered.

Then she nudged Lovie and gestured toward her parents, who were crying on each other's shoulders. "Your mom and dad are taking the loss extremely hard."

"Don't be misled by what you see," said Lovie. "Mike and I were family outcasts. Mike because he was gay and I because I wouldn't live with them after they rejected him."

"How long has it been?"

"It's almost seven years since Mike and I had contact with any of these people."

"Good grief," said Raquel. "My parents know I'm gay. It was rough at first, but now we get along great...unless I ask for an increase in my allowance."

"You were lucky. My dad was the complete opposite. He treated Mike as if he was less than human and Mom sided with Dad like she does all the time." Lovie sneaked a quick glance at her parents. "Jeez, this whole charade makes me wanna puke."

Raquel said, "Some of my relatives gave me a hard time when they heard I was gay, but their attitudes changed when Dad laid down the law."

"What law? Why would they obey your dad? Does he have that kind of authority over them?"

"You could say that," said Raquel.

"You were lucky. We tried using an authority figure to help us," said Lovie. "My family agreed to discuss Mike's homosexuality with our priest. Mike was so excited. He looked forward to the discussion." She smiled as she recalled how Mike reacted when he heard about the meeting.

"We assumed that our priest would have answers if no one else did..." she said. But her voice tapered off into silence.

"...and?" said Raquel.

"It was a horrible mistake. The priest accused Mike of being too weak to resist being gay...and...our father agreed with the priest," said Lovie, crying. It had been a painful episode.

"...and your mom followed suit, didn't she?"

"All the time…that's why Mike and I called her Ditto," said Lovie, drying her eyes with a tissue. "I will never forget how they put Michael out of the house when he needed their love and support more than ever. Fortunately, Michael had an endowment from Chatter."

"…which probably opened like a financial parachute and landed him safely on his feet," said Raquel.

"Of course, but most kids aren't that blessed. Do you have any idea how many destitute gay children are walking the streets every night? They have no place to go and they don't have enough to eat? Most of them became homeless after their snobbish parents threw them out of the house because they thought their kids were gay by choice."

"It's shameful," said Raquel. "My priest told me about it. Have you ever heard of Father Paul, the Plymouth Priest?"

"Who hasn't, he's the priest that's always in trouble with the Church," Lovie replied.

"How right you are," said Raquel sadly. "Anyway, regardless of what people have heard, there is another side to him. He and Dad are childhood friends. He helped my family work our way through my situation and I love him for that."

"It sounds like we could've used *his* help," said Lovie. "Mike was only nineteen at the time. He had no idea why anyone was gay—and especially him, since we were twins and I'm straighter than an arrow."

"What kid in his circumstances wouldn't have questions?" Raquel asked.

"At first, I thought our meeting with the priest was an absolute failure, but now I realize it was the turning point in our lives. That's when Mike founded Gay Activists and started looking for the answer himself."

"…and how did it change you?" Raquel asked.

"After that fiasco, I lost faith in the Church's leadership. Our parents raised us to believe that we could go there and obtain guidance on crucial issues, but this was a travesty."

Lovie, obviously saddened by the experience continued after a brief pause. "Anyway, I was already in college, so I plunged deep into my studies to keep my mind occupied with other things."

Some guys waving their hands at Lovie caught Raquel's attention. "Do you know those fellows over there?" she asked, nodding toward two well-dressed men standing on the opposite side of the grave. But Lovie did not look at them.

"Who are they?" Raquel asked again, while concealing the movement of her lips with her hand.

"They're enemies disguised as cousins."

"What's their crime?"

"We were in high school when I caught them trying to beat the gay out of Mike."

"Oh…how familiar I am with gay bashing," said Raquel.

"The big guy is Louis. I broke his nose," said Lovie, looking straight at him. Then she balled up a tight fist and tapped it on her nose several times to remind him of the day she had given him a judo chop. Louis and his brother turned away. This was not the day to dig up family feuds.

Suddenly, Raquel got excited. "Oh my God, look whose coming. It's Seth! I…I can't believe it."

5. Surprise

The sight of Seth sent blood gushing through the parts of Lovie's anatomy that had been numb since Michael's death. For the first time in over a week, there was a genuine smile on her face. As for Seth, he looked as though he'd lived a harsh life for the past week. He looked a mess. He was unkempt. His jeans and high school jacket was absolutely filthy.

"Seth, where are your glasses? You look as though you haven't slept in days, but I'm glad to see ya. Mike would be pleased that you came," said Lovie, hugging him as if he were Mike.

"This is something I couldn't miss," he said.

"Where on earth have you been? Lovie asked, "We've been worried sick about you."

Seth made no answer. He wasn't ignoring her; he was watching the undertaker place flowers on Mike's casket. He was reminiscing about the good times he and Mike had shared until the scalding hot words of the priest yanked him back to the reality of the burial.

"Michael McGowan was tortured to death by a gang of thugs because he wanted to know why he was gay," said the Priest. "What seemed like a reasonable question cost this young man his life. How could one human being treat the life of another human being with such disregard? What words can I use to explain this? Here we are, in the midst of modern times, but evidently, little has changed since the dawn of man," he said.

The priest shook his head with disgust at man's primitive conduct. "It's been a long time since our ancestors were cave dwellers. Yet, even today, we must remind ourselves that we will remain cavemen as long as we kill each other, as was done here. Let us pray and say our farewells to Michael in silence," he said, bowing his head.

Saying good-bye was the most difficult part of the burial ritual. Lovie, finding no solace or strength in the preacher's words, stood between Raquel and Seth for moral and physical support. She watched the proceedings, but her mind was busy sifting through the scant bits of information she had about the murder. News of Mike's death had come quicker than a flash flood. It hit so hard that it forced her to think about the past two weeks of her life. She thought about everything that she and Mike had done in that time, hoping to uncover a clue to his killer's identity.

The dinner they had shared a few weeks ago crossed her mind. There had been a worried tone in Mike's voice and he kept looking over his shoulder as if expecting someone. She had excused his behavior as normal, since Mike was always hyperactive. But now that he was murdered, she couldn't help but wonder if there was a connection between his nervousness that evening and his death.

Lovie gave Seth a stern look. "Seth, something doesn't add up. Last week, Mike mentioned that you guys were gathering truckloads of incriminating evidence against some crooks. Now, a week later, I'm standing over his grave. What's up, Seth? What can you tell me about those crooks?"

When Seth failed to answer, she grabbed the collar of his jacket and brought her fists together beneath his chin as if she were going to punch him. "Do you know anyone with a reason to hurt Michael?" she demanded. Seth knew Lovie was a capable fighter from the stories Mike had shared; he knew he was in the midst of danger.

Raquel rushed to Seth's rescue. She placed her hands on top of Lovie's and whispered, "I've told you before…your desire for revenge is pulling your strings as if you were a puppet. This is not the way to handle this. Don't look now, but your cousins are hoping you create a scene that embarrasses Mike. Don't give 'em the pleasure."

Lovie gave her cousins a sneaky sideway glance and slowly relaxed her grip. Then she let go of Seth and stepped aside.

Raquel politely straightened out Seth's wrinkled collar, carefully patting it back in place as best she could. Then she whispered a question to Seth too quietly for the cousins to hear.

"What about that conspiracy theory Michael discussed on his Internet show? Is there an evil organization…orchestrating a worldwide conspiracy against gay people? I need a yes or a no."

Seth was so nervous, his entire body was shaking. He looked to see if anyone was close enough to eavesdrop. Then he pulled Lovie and Raquel away from the gravesite so they could talk in private. "We…ah…pissed off some nasty people," he said. "They're called the Inquiry, and the less the two of you know about them, the safer you'll be."

"It's too late for safety Mike is already dead," said Lovie. "…and you know darn well he made me promise to use my doctoral thesis to continue his investigation, don't you?"

"Sure," said Seth, "but you gotta realize that Mike asked you long before all this happened. Be real. If he were alive today, he wouldn't ask that of you."

Raquel said, "But you and I both know that Lovie is going to jump into the fray anyway."

"Unfortunately, yes, I know that," said Seth. "...Lovie, you've got to drop your investigation or they will kill you, for sure."

"Thanks for the advice, but how do I find the Inquiry?" Lovie asked.

"You don't," said Seth. "All you need to do is piss 'em off. At least that's what Mike did."

"I want to know why some people are gay the same a Michael did! So have the new host of Mike's Internet show announce that I'm continuing his pursuit for an answer," said Lovie.

"I'll do it but that's a mistake," said Seth.

"I hear you but have the host rally the audience for me. He should tell them 'The Investigation Lives' and my thesis will reveal the answers to Mike's questions."

"It seems as though the Inquiry didn't want Mike to find the answer," said Raquel.

"You are asking for trouble," Seth warned them.

"We hear you, but tell the host that I will make a guest appearance after my thesis is published," said Lovie.

"Consider it done," said Seth, giving her hand a fidgety squeeze.

"Seth, are you all right? " Lovie asked but got no answer. She could tell that he was frightened by the way he looked over his shoulder as Mike had the previous week.

"What is this with you and Michael? You are frightening me with all this looking over your shoulder stuff. What's going on, Seth? Does this have anything to do with the Inquiry?" She tugged on the sleeve of his jacket.

"I have to go now," said Seth, pulling his arm free from her grip.

"I'll call as soon as I can." Then he fled across the cemetery, hurdling burial plots until he reached the farthest side. There he jumped a fence, got into a car, and drove away.

After watching Seth's behavior, Raquel leaned toward Lovie and whispered, "He ran away like a frightened rabbit. I'm going to follow him." Then she blew a farewell kiss toward Mike's casket and hurried to her car.

Lovie thought about Raquel as she watched her speed away. *"Suspicious" should have been Raquel's middle name; it fits her personality to the nth degree.* Lovie had hired Raquel because Mike suggested she would be a great person to research and validate information, and he was right.

Raquel was secretive when it came to her family, but based upon the small things she had said over the years, Lovie concluded that Raquel's dad was associated with the Mafia. Lovie's curiosity had led her to perform a background check on Raquel. That's when she learned that Raquel had dropped out of Plymouth's Medical Academy after an alleged affair involving another female student. Raquel had fled Plymouth rather than subject her family to a media feeding frenzy. Lovie had decided to respect Raquel's sexual orientation as long as respect flowed in both directions.

After waving good-bye, Lovie headed back to the gravesite—only to see her estranged parents making a beeline toward her.

6. A Family Truce

Lovie and her parents stopped two feet short of a collision. Her mother clung to her dad's arm while waiting patiently for him to give the speech he had prepared for this moment. "I apologize. I acted as though Mike's problem was his fault. I was a jackass, and I hope you can forgive me," he said.

"I hope so, too," said Lovie, after a noticeable pause.

"This is an awful time to confess my error," said Dad, "but, since I don't know if I'll ever see you again, I figured this might be my last chance to apologize—"

Mom interrupted. "There'll never be a *good* time, but this is the *right* time to start mending our family," she said. Lovie was surprised to hear her mom show some independent thinking.

Then Mom surprised her again with a big, comfortable hug. The embrace was something both women had yearned to feel. They clung to each other for a long time, mixing their sad tears for Mike with the joyful tears of a mother's and daughter's effort to reconcile.

Dad watched the two women in his life share their sorrow with a hug. He wanted to join them, but he couldn't get past his guilt and his failure to lead his family during a crisis. "I wish I knew who killed my son so I can return the favor," he said.

"Those are my sentiments exactly," said Lovie.

"Well, I'll be darned," said Dad. "This is the first time we've agreed on anything in years."

"...That's progress. And there's more to come," said Mom. "I saw your friends drive away. Where are you going when you leave here?"

"My limo driver is taking me back to the Marina," said Lovie.

"Pardon me, dear, but do you really think *this* is an evening to spend alone?" Mom asked.

"You should be with family," said Dad.

"This is something Michael would approve," said Mom.

"Come home with us," Dad pleaded.

"We have home cooking and memories of Mike that can be found nowhere else—many of which you have never seen before," said Mom, pulling Lovie toward the family limo.

7. Michael's Repast

The McGowans' home was nestled in Summervale's most exclusive residential area. A maid opened the front door and, for the first time in seven years, Lovie walked through the arched double doorway of the place she had called home.

A spiral staircase seemed to float to the second floor. New colors, new drapes, and new furniture caused a "wow" moment. Many things had changed, but it still felt like home. The family room was spacious and comfortable; a crowd of relatives had gathered there. Lovie took a position across from the wall-mounted, theater-size monitor that displayed the world news with the audio muted.

Lovie's return from extended absence made her the attraction of the evening. One after another, family members came to her and extended warm welcomes. Her aunts and uncles were all smiles and pretty faces that seemed delighted to see her.

"What's happening on the television?" someone asked.

Dad shouted, "That's Father Paul, the priest on trial for allegedly challenging the fundamental teachings of the Church! Hey, everybody, I'm going to turn the sound up for a moment. I gotta hear what's happening to him!"

Father Paul was a devout catholic priest who always carried with him a Bible and a vibrant sense of humor that made him extremely popular with the downtrodden. He was five feet eight, slightly bald with thick sideburns. He had a pudgy pot-belly but he was fast on his feet. Although, only fifty-two his outspoken advocacy against priest abusing children made him an irritation to the Church.

He and his contingent of bodyguards were having difficulties weaving their way through a frenzied crowd of admirers and reporters. A young scrappy female-reporter was scuffling with veteran reporters trying to get an interview with the priest before he entered the Vatican. In a rush to succeed the youngster shoved her microphone in front of Father Paul's mouth,

"Any last words?" she asked, but after realizing how her question could be misinterpreted, she apologized. "Oh, I'm sorry father I didn't mean it *that* way. I wanted to get a few words from you, not your last words," she said tearfully, while the veterans reporters laughed at her.

The youngster's predicament persuaded Father Paul to stop and speak with her, "I'll allow this young lady one question," he announced. Then he turned to her. "Young lady, do you realize it was my talking to the press that put me in this tub of hot water," he said, with a pinch of his well-loved humor?"

The crowd laughed and the young reporter asked her question, "What do you say to those who say priest should not voice public disagreement with the Church?"

"I agree with my critics one hundred percent on matters pertaining to Church policies."

"Then why did you go on TV and express an opinion about priest abusing children that differed with the Church?" she asked.

"Child molestation is a legal issue. All of us including the Church are obligated to obey the law. Thanks for your question," he said, turning to leave. But the young lady was persistent.

"Do you have any closing remarks, father?" she asked.

"Pedophile priests are predators lurking in our Church for young victims. To make matters worse the Church's culture of denial and secretly reassigning serial pedophiles to other parishes places more children in danger. It horrifies all of us to discover egregious cases of child molestation that should have stopped decades ago. So regardless to the outcome of my trial, I will continue to be an advocate for the children and I'll continue to ask the Church to implement my '3-P' program.

"Can you summarize your program for our viewers?" She asked.

"Certainly: whenever abuse by a priest is suspected the Church must notify the *Parents*, the *Police* and the *Papal* hierarchy," said Fr. Paul. For the sake of our children I urge all of you to adopt similar programs in your schools, colleges, in your homes or wherever it fits. You know the difference between right and wrong. Now do something about it…that's what Christ did. Then he blew a kiss to his supporters and walked into the Vatican to face his accusers.

"Wow," said Dad. "He's awesome, isn't he? That guy is about to get his head chopped off, yet he finds time to worry about someone else."

"My friend, Raquel, is from Plymouth and she says the entire City loves him. I can see why," said Lovie, feeling very proud of Father Paul.

"That's the second thing we've agreed on tonight," said Dad.

"You have to like that priest," said Mom. "He's not a rubber stamp for anyone, including the Vatican. We need more people brave enough to stand up and speak."

"Yeah, and he's right about one other thing," said Louis. "Gay priests are attacking our children, and the Church is hiding them."

"Whoa…wait a minute. We need to clear up something," said Lovie, rushing to prevent gays from being slurred. "Pedophiles are the ones who attack our children and they can be either gay or straight people."

A family member said, "I don't care what *you* call 'em. I call 'em 'perverts' and the church shouldn't give sanctuary to them."

Lovie was about to reply when she noticed Mom standing on the steps gesturing for her attention. "I've got something to show you!" She yelled, signaling for Lovie who rushed across the room.

"How did you know I wanted to escape that discussion?" Lovie asked as soon as she was close enough to whisper. Then they hugged each other all the way up to the second floor.

8. Nostalgia

The hallway leading to the bedrooms was wide enough for them to walk side by side and appreciate the family's collection of expensive abstract paintings mounted on the walls.

"Oh, this is Mike's room," said Lovie stopping at a door. Not knowing what to expect she opened it ever so slowly. It was dark inside until she turned on the lights. Then *Wham!* Suddenly, there were hundreds of pictures of her and Mike everywhere. They were in all shapes and sizes. Some were in individual frames and some hung on the walls in collages.

"Mom, this room is a shrine. But it's beautiful."

"This is where I come when I want to be with my babies," said Mom, walking around the room.

"The pictures are marvelous," said Lovie, who couldn't decide which one to pick up first. Then suddenly a smile signaled that she had decided.

"This is my favorite," she said, picking up a baby picture of Mike and herself wearing diapers.

"Mines also," said Mom, gleaming. "Look at those fat-wobbly legs. You kids could barely stand. Why, you'd be on the floor right now if that couch wasn't holding you up," she said.

"It looks like we're having a great time laughing and talking," said Lovie.

"All you are saying is gobbledygook to each other. No one knew what you were saying except the two of you. Do you remember any of that baby talk?"

"Mom that was a long time ago," said Lovie.

"I have a video of you and Mike talking up a storm…I've never heard so much gibberish in my life. If you ever want to take a baby-language refresher, I've got just the video for ya."

"Who talked the most?"

"…you, of course. You were the alpha dog of the litter. If you didn't get your way, you would take Mike by the hand and lead him where you wanted him to go.

"…aw mom, did I really?"

"You were the arm-twister, my dear," said Mom, as she picked up a picture of Lovie standing between two referees. One was raising Lovie's arm high in the air and the other was holding her championship belt for hand-to-hand combat fighting.

"Do you remember this?" Mom asked, handing the picture to Lovie.

"Wow, was I ever that slim?"

"Not only were you slim…you were the Slippery Eel, remember? That's what they called you didn't they?"

"I remember. But I've gained weight since then."

"You've gained weight in the right places…and in the right proportions," said Mom, admiring Lovie's reflection in the mirror.

"I never fully understood why you hired Master Lin to teach me how to fight, of all things," said Lovie.

Mom was momentarily silent, but then confessed, "I didn't want you to be a meek woman, like me," she said.

"Aw, Mom...you were great," said Lovie embracing her mother.

"No, no, I was not! You and Mike didn't like my personality either. I know a lot more than you think I know, young lady."

"Like what?"

"I know you kids called me 'Ditto'. I didn't want you to be a 'ditto,' so I hired Master Lin," she explained.

"Aw, Mom, I'm so sorry," said Lovie embarrassed and surprised that mom knew so much.

"Don't worry about it. You more than made it up to me when you began to bubble with the assertion I lacked. Take a look at these," she said, pointing to several trophies surrounding a picture of Master Lin.

"Thanks for everything, Mom," said Lovie as she picked up the picture. Master Lin was elderly, grey-haired, and known for having tremendous personal discipline. He had not signed the picture, but his motto was on it in Mandarin. Most could not read it. Translated, it reminded Lovie about the importance of mental preparation.

"I thought I was going to be doing a lot of Wushu kicks right off, but there was a lot of science involved. Heck, I had to study anatomy and the roles of our internal organs while my girlfriends were out on dates."

"...but I could tell you enjoyed the training."

"...and what about the training I taught you? Lovie asked. "Do you remember what I taught you to do in case someone tried to mug you?"

"Sure, how could I forget something you taught me on Mother's Day? And you were correct. Gripping the pressure points right here in the wrist really confuses the brain."

"Mom, were you mugged?"

"Me? Heavens, no! I tried it on your Aunt Nancy last year. I grabbed her arm and applied the pressure the way you taught me, and wow she was out cold quicker than you could imagine."

"Oh, Mom, nah you didn't," said Lovie laughing and enjoying the thought of it.

"Sure, I did. But you never should have stopped fighting, young lady."

"I quickly learned that it was not in my best interest to keep throwing people around."

"Yeah, but the Slippery Eel was exciting in the ring," said Mom. "Do you remember when the referee stopped the bout to see if you had oil on ya? I knew he wasn't going to find any, but it was so funny. Wow, what good times we had."

"Oh, my goodness, will you look at this?" Lovie asked as she picked up a picture of Mike. He's was five years old and dressed in a new cowboy outfit. He was standing in front of the Christmas tree wearing boots, a ten-gallon hat with his sheriff's badge pinned to it. He wore a poncho and he had both of his cap pistols firing at whoever was snapping his picture.

"I can hear him yelling 'Pow, pow!'" said Lovie, but her imitation of Mike drew tears from her mom.

"Those were the fondest days of his life…he always said good-bye that way, except last week when he was murdered. Those dirty bastards cut off my baby's fingers—"

"Oh…mom, we've got to find closure and—," Lovie was unable to finish what she was about to say because she was interrupted by the sound of caustic jokes and loud laughter coming from downstairs.

"Shh…" she said, placing a finger over her lips as she tiptoed out of the room and into the hall. When she reached the banister she leaned over it and listened to her cousins ridicule her and Mike,

"…I hear that Lovie goes around picking fights with anyone she suspects of being antigay, so I guess we better duck the next time we see her," said one of them.

"You know all about the repercussions for not ducking, don't you, Louis?"

"…heck, Lovie was such a tomboy in high school, I expected her to turn into a lesbian any minute," said Louis. All the others cousins laughed.

Suddenly, an angry voice came screeching from the stairway. It was Mom, and she was pissed. She came down the steps, pointing them toward the nearest exit. She hollered, "Louis—you and the rest of you heathens—get the hell out of my home…and don't come back!"

It took all of Lovie's willpower to maintain her self-control. She was hurt and ran out of the house crying. She surprised her dad by running past him while he was saying good-bye to some friends.

"Honey, where are you going?" he asked, but Lovie jumped into her limo.

"Take me back to the cemetery. I've got to speak with Michael," she sobbed.

9. The Vow

Lovie stood over Michael's grave, seething with anger toward all those who had hurt him and herself. She wondered if the birth of gays offended God and whether God was with Michael when those bastards had tortured him. *In any event, my thesis will address those kinds of questions!* She swore to herself.

She had difficulty finding the right words to say to Mike until she realized she was still holding his cowboy picture. "Oh my goodness, hello brother," she said admiring his boyish look.

"As usual, I see you're here with me when I need you most." She carefully placed his picture into the grasp of a wreath of flowers. "I came back to tell you that I intend to keep my promise. The Inquiry expects me to come charging after them like a brainless bull, but I'm gonna get them after I have enough evidence to send them to the gallows."

Clenching her fist, she looked up at a starless sky and made her intentions abundantly clear. "My words are for the cowards who killed my brother. You and your kind have humiliated families like mine with mockery and now murder. But starting right now, our roles have changed. I don't know who or where you are, and it really doesn't matter. What matters is this. I'm coming after you and I promise to bring hell with me."

10. Summervale University

Known for its excellence, Summervale University offered a challenging curricula and a supportive environment for achievement. Every doctoral candidate in psychology was required to present a thesis synopsis to the class.

Today was Lovie's turn and she opened her session with a rousing question that snapped the class to attention, "Does it seem as though there are more gays around than ever before?"

The question slammed the class so hard it staggered them like a sucker punch from a heavyweight fighter. Caught completely off-guard, the students sat mesmerized in their chairs as if waiting for the follow-up punch.

"You will be glad to know that my statistical records say it is not your imagination," she said. "Sure, more gays are coming out of the closet, but it's their increasing birth rate you should worry about."

Bewildered students looked from one to another as her unsettling words shoveled all other thoughts from the forefront of their minds.

"What I say is true and if you find this upsetting, wait until you hear about the adjustments you'll have to make." Now that she had their full-attention, Lovie took a position closer to the students sitting in the front row. Then she whispered as if she were revealing a closely guarded secret. "Next year...

and for the foreseeable years to come, the birth rate of gays will be higher than that of straights."

"...higher than ours?" one classmate asked another student, in a low voice that carried a lot further than she expected. "Oh my, gosh," she exclaimed, so embarrassed that she laid her head on the desk and hid her face.

Lovie sensed the class's uneasiness, but she was just warming up. "Ever since I was a little girl, my religion taught me to believe that God has a hand in everything, and I am a true believer. Therefore, since gays exist, they must be here with God's approval."

The class reluctantly agreed with her logic and she continued by asking a question, "If gay people are here with God's approval—tell me why so many religions and governments persecute them?"

Apparently, her words struck home. Her classmates appeared to be deep in thought and nodded in agreement with their internal voices of reason. No one attempted to answer the question, so Lovie went on,

"My thesis asks the kind of taboo questions my brother Michael asked world leaders during his Internet Interviews."

A student stood and said, "I always enjoyed Mike's show. I'm sorry he's gone." Lovie acknowledged him with a nod and continued speaking, "Mike's investigation still lives within me. My goal is to find answers to questions families with gay challenges have been asking for centuries. Then I'm going to publish them in my thesis. So now if you have any questions I will answer them."

The class applauded. One student known to be gay stood and yelled, "Brava!"

Another student said, "What's the big fuss? Gays are freaks of nature. Let's face it; deviations are bound to happen in any baby-making, mass-production line."

"You imply that gays are something like 'lemon' automobiles from an assembly line or maybe you think they are quirks of nature?" Lovie asked.

"Yeah, something similar to that," said the student, nodding his head and hoping to get a measure of student support, but it didn't happen.

Lovie said, "You may not have noticed, but those 'quirks' are now an epidemic. So here's a math problem for you to consider. How many gays do you need to see before you recognize that it's a trend?"

Another student said, "Our religious leaders are handling this, and they seem to know what they're doing."

Lovie said, "The position of some religious leaders regarding gays is equivalent to saying that 'God' made a mistake. So I ask you, by what authority can any of them question the Almighty?"

Another student said, "The Catholic Church has taken a firm antigay stance."

"I'm Catholic, but I don't like the leadership," said Lovie. "I most certainly do not want to offend the Holy Church, but the leaders give the impression that they condemn gays and condone pedophiles. To be honest, I find this very confusing."

A female student with a meek voice said, "I don't think our Church condone priest abusing children, do you?"

"Maybe and maybe not," said Lovie. "It's what they do that counts. You are aware of the laws God gave us in the form of the Ten Commandments?"

"Of course," said the woman.

"Then you understand that the commandment, 'Thou Shall Not Steal'... prohibits stealing a child's innocence," said Lovie.

The question shocked everyone in the classroom but the young lady recovered quickly, "It's a sin no matter how we look at it," she said and Lovie agreed with her.

The classroom was silent until another classmate broke the ice. "—Lovie, you should think about this because you're committing political suicide."

"I apologize if my discussion of taboo subjects offends public moralities and established customs but I'm compelled to seek out the truth and share it with anyone brave enough to read about it," she replied.

"Then, maybe you should pick a topic easier for the faculty to support."

"I can't do that. Taboo issues are all around us," said Lovie. "Our society has an urgent need for someone to address the issues concerning homosexuals, pedophiles and bullying. Personally, I can wait…but the children walking the halls of our schools cannot."

A male student big enough to be a NFL lineman stood and gave Lovie a salute. "You're not using sticks and stones; you're using common sense and that makes you a fearless warrior," he said. Then he left for his next class.

Since this was Lovie's last class of the day, she went home as well.

11. A Bad Hair Day

Raquel was in Lovie's Condo analyzing data she had collected from a survey. The day had not gone well and it got worse during her lunch break. She had been watching TV when a Vatican spokesman moved her to tears when he released some heart-breaking news about Father Paul.

Clickity- clack went Lovie's metallic key as it entered the front door latch. Raquel quickly dried her eyes and hid her sadness behind a smile as Lovie walked in.

"Welcome home, girlfriend," said Raquel, who expected Lovie to be high-spirited. Instead, she had a serious case of sad-face and her mind was somewhere else.

"Why are you wearing that long, slacked-jawed face? It doesn't become you at all," Raquel asked. But Lovie's silence was her only answer. Her usually rosy cheeks were pale and her bright eyes and smile were missing.

"Uh-oh...it must have been your presentation," said Raquel. "Obviously, it didn't go according to plan. Tell me what happened," she said, trying to pry a reply from Lovie.

"Blah, blah, and blah are the three best words to describe what happened. My classmates were stunned," said Lovie.

"Then the jury is still out?" said Raquel.

"Oh yeah, it's out all right. They're in a coma and if I'm lucky they'll wake up in a few years."

"It was that bad?" Raquel asked.

"We'll have to wait and see," said Lovie, quickly changing the subject. "Let's talk about Seth. Where did he go after he left the cemetery?"

"That's one weird kid," said Raquel, sipping her iced tea. "I followed him to the warehouse district. He entered a vacant building…where I suspect he lives."

"Then what happened?"

"I waited several hours, but he never came out. Maybe you should avoid him for a while."

"He's weird, but he's a good kid," said Lovie, moving closer to the TV screen to get a better look at the picture paused on it.

"Isn't that Father Paul standing between those two big guys? What's going on?" she asked.

"They're his bodyguards," said Raquel.

"How do you know that?"

"Dad's been providing protection for him ever since his life was threatened a few years ago."

"What is that name you Plymouth people call him?"

Raquel's admiration for Father Paul shaped the prettiest smile on her face. "He's 'the Plymouth Priest,'" she said. Then suddenly, anger erupted and erased her smile. "He is also a pistol that shoots off comments that get him in trouble," she added.

"What's he done now?" Lovie asked.

"I would strangle that man if all of Plymouth didn't love him so much. Can you believe he criticized the Vatican's policy toward gays while he's on trial for accusing the Church of harboring pedophile priests?"

"Gee, what was he thinking?" Lovie asked.

"He was thinking about the victims at the cost of his personal safety," said Raquel. "He has to be more strategic...but it's been that kind of day. He struck out and so did we."

"Oh my, gosh, go ahead and poop on me. Everyone else is doing it," said Lovie. She flopped down on the sofa and curled her legs beneath herself, as if surrendering.

"I thought we were making progress until I got caught between the contradictions of two reports," said Raquel, who closed the news and opened her research program on the same screen. Then she used her fingertips to enlarge and partition the touchscreen so she could display two reports side by side, for comparison.

"These reports are from our survey. The report on the left is from the US Bureau of Vital Statistics. They say that gays represent four percent of the world's population. The report on the right is from a website called 'Gay Facts' which says the percentage is closer to *twenty*-four."

"That's a huge difference," said Lovie.

"Tell me about it. That's why, yesterday, I called and spoke with both sites."

"...and..." said Lovie, urging her to continue.

"Vital Stats was not helpful at all. But Gay Facts offered to prove their numbers. They also advised me not to believe the government."

Bingo...Raquel's words rang true with Lovie because Michael had given her the same warning. "What about Michael's question? He wanted to know why some persons are gay. So we used 'Question Ten' to establish a criterion for any explanation vying for acceptance as the truth. What kind of results did we get?" Lovie asked.

"I think you are going to like this," said Raquel as she displayed the criteria established by those who responded to the survey:

1. The 'truth' must apply to every man and woman without exception.

2. The 'truth' must explain the trend that's increasing the number of gay births each year.

3. The 'truth' must explain to our satisfaction why we have bisexuals, gays, hermaphrodites and transvestites.

4. In the case of multiple births, the 'truth' must be capable of affecting one child and not necessarily all the others.

"That's a tall order but we can't accept less," said Lovie.

"That's not all. Today, Gay Facts and three other websites that helped us... disappeared."

"What do you mean...disappeared?"

"They're gone, kaput, they've shut down," said Raquel "...and it wasn't by magic."

"Do you recall the names of the websites?" Lovie asked, rubbing her stomach because her guts usually gnawed at her when she sensed a threat.

"Sure, I've got the names right here," said Raquel, scrolling through the pages displayed on the monitor. "Ah…here they are. 'Lesbian Lifestyle' is gone. 'Gays' is gone. And my favorite website, 'Mistreated,' disappeared as well," said Raquel.

"Why am I not surprised? Did you know they were websites highly recommended by Michael?"

"I suspect we are being steered away from the truth," said Raquel. "Do you trust the remaining websites?" she asked.

"Not as much as I did this morning," said Lovie. "I'm edgy, and my personal antennae are searching for signs of danger."

Raquel began bouncing up and down on her toes like an athlete limbering before a big game, "I'm nervous also, but that's how I get my motor started. I feel my fighting blood warming up. "Vroom, vroom," she said trying to sound like a NASCAR driver gunning her motor before the big race.

12. Saylor

Later that evening, Lovie and Raquel dined at the Leprechaun, an upscale seafood eatery famous for its rock lobsters and oysters on the half shell. During cocktails, an odd-looking stranger walked up to their table and introduced himself. "Pardon the intrusion, ladies. My name is Saylor. I was a friend of Michael and I'm sorry I missed his funeral."

"How did you know my brother?" Lovie asked, while trying to ignore Raquel kicking her beneath the table because the stranger looked highly suspicious to her.

Saylor said, "I joined Gay Activists after watching one of Mike's Internet interviews. Shortly afterward, he and I began working together. You must be Raquel," he said with a curtsy.

"That's me," said Raquel. Her suspicion was bursting her seams while she tried to figure out his country of origin. Saylor most certainly was an odd looking fellow. He wore sunglasses with a frame large enough to hide most of his facial features. His attire resembled the style made popular by India's Prime Minister Nehru, years ago, except that he wore a turban that covered his ears. He had an Asian complexion and his skin was unblemished. He had neither eyebrows nor eyelashes. In fact, if he had any hair at all, it was not apparent.

"Mike and I exchanged useful information," he said. "He told me about your promise to continue his investigation. If that's your intention, I'm here to warn you—"

"...about the Inquiry?" Lovie interrupted.

"They killed Mike, and they will kill anyone else who gets in their way," he said.

"We need evidence to bring them to justice. Do you know where I can find them?" Raquel asked.

"They're evasive. Several years ago, they ran a medical research facility on an island."

"How would you know that?" Raquel snapped.

"I'm a merchant marine. My ship delivered cargo to them until they were booted off the island."

"What kind of medical research did they do?" Raquel asked.

Saylor said, "Mother Nature had an idea of where she wanted to take humankind, but the Inquiry had other plans."

"...and you know this because of...what?" Lovie asked.

"I've read the Inquiry's notes and I've seen the Newbies," he said.

"...Newbies...what do you mean...Newbies?" the women asked in unison.

Surprise was in Saylor's voice when he spoke. "So Michael didn't tell you about them?" he asked.

"What's a Newbie?" Lovie asked.

"They're all the evidence you'll need to convict the Inquiry."

"Has anyone seen them besides you?" Raquel asked.

"The science team that replaced the Inquiry and, of course, the Zoo Keepers saw the Newbies every day before they became adults."

"...Zoo Keepers?" Lovie asked. "Who are they?"

"That's the offensive name some straights used to describe those responsible for educating the Newbie babies when they first arrived," said Saylor.

"Arrived? Arrived from where? And where is this island? Is there a number I can call?" Lovie asked as she turned on her PenSet to add the number to her contact list.

"We want to meet the Newbies," said Raquel.

"The only way anyone can see a Newbie is to sneak onto the island."

"Okay, now we are back to Newbie Island...where is it?" Lovie asked.

"I came here to warn you *not* to get involved, so that's the last thing I'm telling you," said Saylor. "But maybe I'll answer your question when you girls know more about what you're up against."

He gave Lovie a small piece of paper with some scribbling on it. "That's how to reach me if you have an emergency," said Saylor, who bowed good-bye and went his way.

13. The call from Seth

The next morning, Lovie got out of bed before the alarm clock woke her up. She had a full schedule so she'd planned today's chores and errands down to the details. She was already behind schedule when the phone rang. She stared at it and considered not answering, but it was a good thing she did.

The sound of Seth's voice was a surprise, but she recovered quickly. "Are you still jumping over graves?" she kidded him.

"It won't matter, because the next one will have my name on it," Seth replied, as though his demise were a foregone conclusion, in the hands of someone else.

"Seth Bluestein, I want you to hear me clearly…you will stop fighting when I stop fighting, and I'm not ready to stop," she demanded.

"They killed Mike to get what I got and…I'm gonna get the same," he said.

"What are you talking about?"

"I've got their book. I'd like to show it to ya before I give it to Lieutenant Muggs," he said. Then he began talking to himself. He rambled and answered his own questions as if Lovie were not on the phone.

"Seth…Seth!" she yelled, trying to get his attention, but he was too engaged in talking to himself to hear her.

Lovie heard him say, "The Inquiry has too much influence in high places. I can't trust anyone…anymore," he said, in a low mumble.

"You can trust me, can't you? Surely, you remember that?" Lovie's question must have struck a friendly chord that reminded Seth that he was on the phone with her.

"I'm sorry, Lovie, were you gonna say something?" he asked

"Where are you, Seth?"

"I'm sure Raquel told you where she followed me," he said.

"I can be there in an hour," said Lovie checking the time indicated on her PenSet.

"All right, but you can't mention our meeting to anyone. I mean it. Not even Raquel. She's our friend, but even she could have a slip of the tongue that'd get me killed."

"You have my word." Lovie gathered her keys and purse and headed for the door as quickly as she could.

"Oh, oh…Lovie, would you bring a couple of cheeseburgers? Put everything on them—and a large root beer float, will you?"

"I'll be there in a flash," she said rushing out the door.

14. The Rendezvous

Lovie showed up at Dockside's Warehouse, but Seth was nowhere in sight. The warehouse looked like an airplane hangar—it had a half-moon shape with two large front doors that slid open on rusty tracks. It took Lovie's best effort to slide one of the heavy doors open enough for her to squeeze through.

The blast of sunlight that came through the opening frightened the seagulls in the rafters and revealed a warehouse cluttered with junk.

Stepping over debris, she yelled, "Seth…hey, Seth? It's Lovie. Where are you?"

She got no answer, so she ventured farther inside. The warehouse was spacious but junky. The lights were off, but it wasn't completely dark. There were rows of windows near the ceiling. The sight of crumpled junk food wrappers and empty food cans caught Lovie's attention. She felt relieved when she spotted Seth's high school jacket folded over a makeshift bed. It wasn't much, but it assured her she was in the right place.

The sheer size of the warehouse made her feel small and vulnerable. It was like the intimidation she felt whenever she fought an opponent in the ring who was bigger than she was.

She'd heard seagulls since entering the building, but now they were too loud to ignore. The birds were screeching angry noises as if they were fighting over food. Her eyes followed the noise. That's when she saw Seth, crucified between two windows. The birds were fighting over his flesh.

The horrific sight pumped terror and panic through her veins. She dropped Seth's hamburgers and ran recklessly toward the exit until she tripped over a rubber tire and slid across the greasy floor. The fall hurt, but it added to her determination to get out of there alive. She stood and sprinted toward the door. Ahead she saw a shaft of sunlight shining through the opening and she ran straight through it like a blur...r.

15. FedEx

Lovie burst through the front door of her Condo huffing, puffing and darn near out of breadth. She was still gasping for air when she slammed on the deadbolt and rested her back against the door.

"My, gosh, girlfriend what's wrong?" Raquel asked rushing to her rescue.

"They killed Seth," said Lovie her face tear-stained and glowering.

"What…?"

"My, gosh, he was a bloody mess," she sobbed.

Raquel eyes enlarged as she struggled to make sense of the Inquiry's ruthlessness. "Why…why torture a child? He was barely twenty, for heaven's sake."

"Seth and I spoke earlier this morning. He said he had a book that belonged to the Inquiry and they wanted it back."

"Did he say he was sending you anything?"

"He asked me to come see the book. That's why I went there. Why did you ask that?"

"FedEx dropped something off a few hours ago. I don't recognize the name of the sender, but I'd swear that's Seth's handwriting." Raquel retrieved the package. "It's got some weight to it," she said, bouncing it up and down to judge the weight. "Two pounds, at least," she said, tossing it to Lovie.

"Gimme the scissors," said Lovie.

"I'll do better than that. I'll cut it open for you." Raquel sliced off the end of the package and turned it upside down. The contents slid out of the wrapping and onto the couch.

"It looks like we have a leather-bound book with the name 'Cyclops' embroidered on the front cover," said Lovie rubbing her hands across the cover to feel the stitches.

"Mmm...and a flash memory cartridge is taped to it," said Raquel.

"...and that's where we'll start," said Lovie, turning on her PenSet.

She aimed the flash memory cartridge at her PenSet and pressed the '*send*' icon. A request for her password appeared on the monitor. After she complied, the image of Seth appeared and he began talking to them. "Hello, ladies. The Inquiry almost caught me a while ago...I'm all right, but my narrow escape convinced me to send their book to you right away. I've added a video I made for Mike when he didn't show up for our meeting."

Then Seth pointed his finger at them and said, "There's enough evidence in the logbook and on the video to send the Inquiry to the gallows. Or it will send both of you to your graves, if you are not careful."

"Oh, my-gosh," said Lovie. "He must be talking about the last meeting marked on Mike's calendar."

Excited, both ladies sat and watched Seth and a guy twice his size sit down in a booth. He was a big man with the square jaw and broad shoulders of a Marines drill sergeant. You could tell that at one time, he'd been a giant of a man. But now he was grey-haired, worn by time, and with eyes that begged forgiveness.

"I'm Winston," he said, reaching across the table to shake hands with Seth. "I recognize you from the Internet show. Where's Mike? He's supposed to be here." Winston looked around the pub, hoping to see Mike.

"Mike will be here shortly. Let's order something while we wait," said Seth. And to a waitress walking by: "Two beers."

Winston yelled, "Make it three drafts in cold mugs if you will, my lass!" He whispered to Seth, "I usually order two at time." They drank beer for over an hour. By then, beer suds covered Winston's moustache.

"If I sit in this here booth much longer, someone will spot me. So let's talk about why I'm here," he suggested. "I'm a sailor by trade and from the year 2000 until 2010, I sailed as first mate on a ship owned and operated by the Inquiry. The Cyclops was her name. She was a one-of-a-kind vessel designed to carry a one-of-a-kind cargo."

"And her cargo was…?" Seth asked.

"Slaves,"

The answer caused Seth to choke a bit in disbelief. "You mean enslaved *people*?"

"The Cyclops transported gay people from countries trying to get rid of them and believe me…there were a lot of 'em."

"…and what did the Inquiry do with these gay slaves?"

"We delivered them to Newbie Island. My crew never got farther than the docks, so we don't know what happened after that. But we heard all kinds of rumors."

"Like what?" Seth was anxious to see what the conversation would reveal.

"Well…ah…we heard that the slaves were used in ungodly experiments. It gives me chills to think about it," said Winston.

"Did the gays ever put up a fight or try to jump ship?"

"There was no escape. The Inquiry's medical team kept them highly sedated and they fed the slave intravenously. Mind you that my crew was as humane as we dared be under the circumstances."

"Unless you were one of those slaves," said Seth.

"Oh come off it, mate. I don't need some young pup to judge me," said Winston. "God knows I have a penance to pay for my sins and I'm trying to do that, if you let me."

"Sorry," said Seth. "…but this is hard to grasp."

"Look here, I'm not proud of my doings, but it was the best job I could find. I've wanted to make amends for a long time. When I heard about what you were doing at Gay Activist, I figured this was my chance."

"I'm with you."

"I got something for you," said Winston, sliding the logbook across the table.

"Everything you need to confirm Mike's conspiracy theory is right there." He pointed at the book.

Seth opened the logbook and began browsing the pages while Winston continued. "That's the Cyclops's logbook, in the captain's own handwriting. You can see for yourself that it gives the dates, times, and places where we picked up the slaves." Seth nodded yes as he tried to read and listen simultaneously.

"Did you know that reading that book is the same as listening to the cap'n himself?" Winston asked.

Seth's mind was cross-referencing everything he was hearing to see if the facts supported each other. "You guys transported slaves for ten years and didn't get caught?" he asked somewhat skeptically.

"The Cyclops was fitted with the kind of stealth equipment thought to be limited to US military aircraft. But I tell you mate, if my captain wanted to avoid detection none of the equipment out there would see us."

"That's a good answer," said Seth.

"He was a good captain. But he knew his days were numbered. So one day, when we were on shore leave, he...uh...he gave me that logbook and told me to get as far away as possible."

Seth waited for Winston to continue speaking. And he did, "...you know what else he said?" Winston asked.

"Why don't you tell me," Seth suggested while saluting Winston with his mug of beer.

"Make this book do some good. That was the captain's last order."

Seth said, "It would make the captain happy to know that I'm a gay person with personal reasons to see that his orders are obeyed."

"Aye, aye sir, that's good enough for me," said Winston. He stepped out of the booth and stretched his long arms. "Now it's up to you chaps to put this book to good use. I'm sorry I missed Mike," he said as he drained the last drop from his mug. Then he wiped the suds from his mustache with his shirt sleeve, and walked away.

As soon as Winston disappeared, Seth stood in front of the camera waving his hands. "Hey, Mike, I made this video so you could see where you would have sat had you made this meeting like you promised. You missed it, so I guess even you can be late sometimes." Then Seth imitated Mike's fast draw, saying "Pow, pow! I'll see you in the morning, old buddy."

Lovie said, "Poor Seth. The kid had no way of knowing that Mike was being tortured while he was recording this."

"I've read a few pages," said Raquel, holding the logbook above her head. "It identifies everything except the location of Newbie Island but we have a bigger problem; The Inquiry is coming after this."

"Good grief! I'm sorry for getting you into this mess," said Lovie, "Saylor was right. I had no idea what I was getting us into."

16. Dr. Scorn's Scheme

The Summervale Hotel, known for its extravagance, outdid itself when it came to the penthouse suite. The Inquiry leased the two-story penthouse for Dr. Scorn, and this morning, he was conniving with Konnerman and two of her thugs when the phone chimed. The video caller ID displayed Monsignor Mancini.

Sergio Mancini was a high achiever from birth. He was born to an affluent Italian family who loved and encouraged his pursuit of knowledge. His parents were delighted when he showed an interest in science and the Holy Church. They sent him to the best schools that taught science and Catholicism but they never imagined he would earn doctorial degrees in divinity and biochemistry, but he did.

The monsignor's unique credentials and obedience to the Church earned him an assignment he did not want. His job was to act as liaison to the Inquiry and report their dirty shenanigans back to the Pope and to the Pluralists. It was one of those lowdown-dirty jobs someone had to perform, and he was the best qualified.

Dr. Scorn pressed the *talk* button. "What can I do for you, Father?"

"I'm returning your call. And we need to make this brief," said the monsignor in a condescending voice.

"I called to find out how you're progressing with the Serum Project," said Dr. Scorn.

"Completion would be a lot closer if I had fewer interruptions."

The monsignor's sour attitude irritated Dr. Scorn which prompted him to question Konnerman. "Is it my imagination, or am I detecting pheromones that are giving this conversation a sour taste?"

"It's the latter," said Konnerman.

Dr. Scorn gave the monsignor a mean look. "You're not very talkative today, Father. Maybe you would like to renegotiate our arrangement?"

The monsignor's stared at Scorn with complete mistrust and he made no secret that he expected foul play. Scorn noticed his guarded manner right away. "Don't look so surprised," said Scorn, "Even professional athletes are renegotiating their contracts nowadays."

"What're you up to now?" The monsignor asked.

"Aha, he can speak. Now we're talking. My dear monsignor…need I remind you of the existing arrangement? You know the part that requires the Church to keep a lid on Father Paul's mouth."

"There's a lid for every pot," said the monsignor.

"What do you mean by that?" Dr. Scorn asked.

"I'm saying there's no need to worry; a decision regarding Father Paul is forthcoming. It's the apocalypse that's bothering the Church."

"Go on," said Dr. Scorn.

"Several of my colleagues believe that the path we are travelling will lead to man's demise, for sure," said the monsignor.

"...so you want to tell the people of the world that their days are numbered...is that what you have in mind?"

The monsignor nodded yes, which frustrated Dr. Scorn. "That would start a worldwide panic before you've tested the serum. The Inquiry doesn't like the delay any more than you do, but people are more manageable when they are in the dark."

"Have you considered what you'd say to the World Court when they ask you to explain how you know danger is afoot?" Konnerman asked.

The monsignor looked away from her rather than reply so Scorn offered another question, "Are you going to admit what we did to the Gay slaves as well?"

"I didn't do anything wrong to anyone," said the monsignor defiantly.

"My dear Monsignor, if the truth gets out, they're gonna hang you right along with us," said Scorn, looking at his thugs and they agreed.

"You are not listening," said the monsignor.

"No...you're the one who's not listening," said Dr. Scorn, fiddling with the weapon hidden within his oval ring. After adjusting it a bit, he extended his hand and pointed the ring toward a large jade urn a few inches to the left of the monsignor.

He pressed a trigger near his knuckle and a high-frequency blur streaked across the room and shattered the urn to bits. After making his point, Dr. Scorn said, "Tell me, monsignor, who will continue your work on the serum if something tragic happens to you?"

The monsignor did not reply, nor did he shy away from the implied threat.

"So let's keep it simple. Obey the agreement and you will live long enough to complete your precious serum. Then you can save all the people you want. Do we still have a deal?"

"Scorn, you're a cruel and evil creature. For the sake of humankind, I hope all evidence of your genes become extinct."

"They *are* pesky little devils…now that you mention them," laughed Scorn. Then he turned toward his thugs and said, "It sounds like we have a deal."

"My only deal is with almighty God," said the monsignor. The screen went blank as he terminated the conversation.

Dr. Scorn shook the fingers on both of his hands as if they had touched something scalding hot, "Whew," he said blowing air on his fingertips to imply he was cooling them down.

"Wow, tell me is the monsignor fuming mad or what?" Konnerman asked in jest.

"Who cares? After we take control of the Serum we will sell it to the people and be set for life," said Scorn, rejoicing at the strangle-hold he had on the Church.

Suddenly, Scorn's smile vanished and he growled at Konnerman, "None of this matters if you fail to get my logbook."

"Where do I look?" she asked.

"I'm sure those two bothersome women know of its whereabouts. The log-book can set us free or send us to the gallows. After you get it, we will give those two witches a medieval stake cookout," said Scorn, delighted with his wordplay.

"I can hear their screams now," said Konnerman, bowing and quietly backing her contingent of goons out of the room.

17. Girls on the Run

The thugs burst through the front doors of Lovie's and Raquel's condominium, raring for conquest, but they were too late. The girls had left hours ago with the intention of putting as much distance between themselves and Summervale as possible.

They were driving Raquel's car to Plymouth to live under the protection of her father. Raquel remembered him teaching her that driving a privately owned car made it more difficult for pursuers to find their prey. So there she was behind the wheel. Lovie sat in the passenger's seat watching the Internet news and hoping to learn more about Seth's murder when suddenly an announcer interrupted.

"We're going to leave regular programming and switch to Rome where the Vatican Court is about to announce its decision regarding Father Paul." The TV anchor's desk disappeared and the screen filled with a crowd of protesters carrying signs in support of their beloved Father Paul.

"Put the news on pause while I turn the driving over to eChauffeur," said Raquel. She pressed the activation button on the steering wheel and the software didn't take long to load. "Where would you like to go?" eChauffeur asked.

"The home of Don Carlos Graciano," said Raquel.

"One moment please while I check the fuel cells," said eChauffeur. A few seconds later, it said, "Everything checks out fine. You can relax while I drive."

"That's what I want to hear," said Raquel releasing her grip on the steering wheel, but as a precaution, she kept her hands close to it until she was confident eChauffeur had control of the vehicle. Once satisfied, she spun her seat around to face the monitor. "Now I'm ready," she sighed.

The women watched Monsignor Mancini approach a row of microphones and several eager reporters. Usually, the monsignor appeared calm and collected, but today he had the jitters. "First, I'm pleased to announce that next month, the Holy Church will publish new policies regarding pedophile priest."

The bipartisan crowd greeted the delay with boos but the monsignor continued, "We're also delighted that Father Paul has accepted a rewarding new challenge in the Northern Territory. He will be leaving for his new assignment as soon as he wraps up his affairs in Plymouth."

The monsignor watched the crowd's stunned reaction. They moaned and jeered their disappointment. The flushed pinkish color of his long, narrow nose expressed his own shame as he hurriedly left the podium to avoid questions.

"Darn it! Whose side is that monsignor on?" Raquel asked.

"That's what I've been wondering. He's definitely not on our side," said Lovie, who balled up her fist and tapped her nose as if designating the monsignor as a person not to be trusted.

The Internet news didn't miss a beat in covering this breaking story. As soon as the monsignor left the podium, they switched the television audience from Rome to the City — State of Plymouth.

Father Paul was standing in his churchyard, basking in the cheers of his supporters. The townsfolk were proud of him and eager to hear their native son speak.

"The Church sentenced me to the Northern Territory when they should have taken steps to restore the integrity of the priesthood. Punishing me is all right if that's what it takes for me to remain a priest. But it's not all right for children to continue waiting for adults to treat them with dignity. I ask you, how much discussion is required before a group of highly educated men use the common sense God gave them?"

There was an unusually long pause. Then Father Paul finally admitted, "I'm hesitating because I wonder if I should speak my mind which will undoubtedly get me in more trouble."

A supporter yelled, "Whatcha got to lose, Father? They can't send you any farther North than you're already going." These words of encouragement were precisely what the crowd needed to loosen sorrow's grip on their hearts. Everyone, including Father Paul, laughed at his self-made predicament. Then he spoke his mind.

"Okay…here's what bothering me. Sometimes it feels as though a dark, sinister force is working in the background to influence the church's behavior. I know…I know, it doesn't make sense, and it's an awful thing to say. But sometimes that's what comes over me." Then he shrugged his shoulders as if trying to prevent a chill from invading his body.

"What do you mean?" shouted a supporter.

"I wish I knew," said Father Paul, shaking his head.

"When're you leaving, Father?" someone asked.

"I knew this day was coming. My affairs are in order, so I plan to leave later today."

A woman in the crowd wiped her tears with one hand while holding the hand of a child with the other. She waved and got Father Paul's attention. "We are asked to wait while our children are being molested with indifference. Will your new assignment prevent you from helping us?"

Father Paul said, "Be assured that nothing is going to keep me quiet."

"Damn it, I don't like this turn of events," said Lovie. "I was hoping we'd get a chance to speak with him when we got to Plymouth."

"I know the feeling," said Raquel, "But it's not all bad news."

"Why's that. What do you know that I don't?"

"To get to the Northern Territory, Father Paul has to get a flight out of Anchors International, which is about thirty minutes that-a-way." She pointed.

"Right on," said Lovie. "I feel better already. Let me think…uh…uh…we need to talk with Father Paul without any reporters around. Is there a chance you can arrange that?"

"I'll phone Dad and ask," said Raquel activating a speed-dial option.

"May I help you?" a male with a bass voice asked.

"Cousin Mario, this is Raquel. May I speak with Daddy?"

After a short pause, a different voice intoned, "Hello, Butterfly." It was Carlos Graciano, the mafia don of Plymouth.

"Dad, are you watching the Internet news?"

"With my own eyes, I just saw my own church slam dunk my best friend. Why do you ask?"

"—Father Paul, Lovie, and I all need your help." There was silence while each waited for the other to speak.

"What kind of help?" asked the don who seemed to be testing the water with his toe prior to taking the bait.

"Dad, you once told me that Father Paul did our family a great favor and you have always looked for an opportunity to repay him…trust me, Dad, this is it."

"I'm listening," he said.

"Lovie and I believe we know who's got the stranglehold on the Church. We want to discuss it with Father Paul without cameras and news reporters. Can you arrange it?"

"Where are you?"

"We'll be in Anchors in thirty minutes—which reminds me…we need another favor."

"You need only ask," said the don.

"Lovie and I must remain anonymous, so will you book lodging for us in a low-profile motel? Don't use our names."

"Do you want me to send help?" he asked.

"We're fine. I will call you and Mom after we meet with Father Paul."

"A friend of mine owns the Cliffhanger Inn. It's an out-of-the-way motel on the outskirts of Anchors. I'll register you girls there under your Aunt Marie's name."

Lovie said, "I've made a note of it."

"Now back to this meeting with Father Paul. I'm looking at his flight's ETA…can you visit Anchor's Cathedral this evening? Would seven o'clock be okay?" the don asked.

"We'll be there," said Raquel. She took control of the car from eChauffeur. A few miles later, she pulled off the expressway onto the Anchors exit ramp.

18. A Posse of Thugs

The Inquiry's computer network was checking hotels, airlines, cruise ships, and private charters looking for the girls. They shot blanks until they got word that the eChauffeur in Raquel's car was using satellite navigation to get to Don Graciano's home in Plymouth.

Konnerman and her thugs jumped into their van and pursued them to Plymouth. They drove past the Anchors exit and continued north for several hours before they learned that the girls had exited at Anchors. A frustrated group of uncivilized brutes reversed their direction and headed south.

19. Anchors' Cathedral

The nearby City of Anchors built an international airport to accommodate Summervale's growing commercial needs and both cities flourished as a result. Anchors' was a picturesque old-world place with buildings influenced by Moorish architecture. Saint Joseph's Cathedral was no exception. Built with rough, colorful stones and covered with several gabled roofs, it was a masonry masterpiece amid a bounty of tall palm trees.

The Cathedral sat atop a high hill overlooking the ocean. The church had a spacious, fenced backyard that led to a cliff with a steep drop. Thanks to the fence, Churchgoers could stand safely close to the edge and look down at the piers where parishioners who came by sea docked their vessels.

Although the girls arrived early, Don Graciano had two bodyguards already there. Paulo, a young man who had grown up with Raquel, seemed pleased to see her. He waved and rushed to open the car door for her.

Raquel said, "Paulo, seeing you brings back memories of happier times," as she embraced her childhood friend. A moment later, Father Paul's car pulled up with several more bodyguards.

Raquel was glad to see Father Paul, but she was surprised at how much he had aged. "Oh my...he looks so much older and he looks tired." She waved for Lovie to follow her. Then she ran to Father Paul to exchange greetings.

"Don Carlos said for you to be careful," said Father Paul.

"I promise," said Raquel. "And this is my friend, Lovie. She's been dying to meet you, our hero."

"I hope I don't disappoint you ladies," said Father Paul, inhaling a deep breath of air.

"We have arranged for you to meet over there," said Paulo. He led them into a building adjacent to the church. There he showed them to a conference room with a long, oak table and several comfortable chairs.

"We will be outside if you need us," said Paulo, smiling at Raquel. She blushed when their eyes met.

A wooden bowl of fresh, succulent fruits was in the center of the table. Father Paul took a handful of grapes and passed the fruit bowl to Lovie and Raquel.

"What do you want to talk about?" he asked with a probing stare.

"What can you tell us about the Inquiry?" Raquel asked.

Her question startled Father Paul, who almost swallowed a grape whole. "This is not the conversation I expected," he said. He waited a moment, and then spoke in a quieter voice. "A few years ago, Father Thomas—a trusted friend—warned me to be careful. He told me several hard-to-believe stories implicating the Church and a mysterious group of religious fanatics called the Inquiry."

Father Paul paused and looked around the room. When he was sure no one else was in there with them, he continued to share what he knew. "A week later, Father Thomas disappeared. I looked for him but I ran into a stone wall. Shortly afterward someone broke into my living quarters and tried to smother me. I'm alive today because Don Graciano protects me. So, like everyone else, I don't *know* if the Inquiry exists…but I think it does."

"It exists and we can prove it," said Raquel.

"They killed my brother and his friend," said Lovie.

"You ever hear of the Inquiry shanghaiing gay people and taking them to a remote island?" Raquel asked.

"Heavens no, that's a new one," said Father Paul. "I don't know of it personally, but I know someone who can find out for us."

"Who would that be?" the ladies asked in unison.

"Monsignor Mancini," he said.

"Yikes...isn't he the monsignor who's been trying to wring your neck?" Raquel asked.

"I wasn't planning on asking for his cooperation. Paulo came into possession of the Monsignor's password for the Church's computers. I was thinking we might use it to find the answer to your questions," said Father Paul, folding his hands over his chest and mentally asking for forgiveness in case he needed it.

Lovie set the first two pens from her PenSet on the table. Then she powered them up and issued her instructions: "Give me three mice, one keyboard and one twenty-four by sixty-four inch, adjustable screen."

"Father Paul, connectivity to the world awaits you," said Lovie, pointing to the virtual devices displayed on the table. "I've also given us the option of uploading to my Cloud account anything we think will build a legal case for us."

The table's top was made of four thick planks of smooth wood long enough for the three of them to sit on the same side of the table and view the monitor. Father Paul brought up the Church's website. Then he selected *Contact us* from the banner at the bottom of the page. A moment later, a preaddressed e-mail template appeared on the screen. He replaced the address with the one priests used to interface with the Church. He pressed *send* to bring up the firewall that protected the Church's network from intruders. Then the screen went blank except for a flashing cursor awaiting a password.

"Here goes," said Father Paul as he typed the monsignor's password and pressed *enter*. Suddenly, a webpage he'd never seen before appeared. "There, I've done it," he said. "We're inside the Church's network. Now one of you needs to take over...I'm not very good with this technical stuff."

"I'll do the typing and you guys give the orders," said Raquel, positioning the virtual keyboard in front of herself, "Now How do we find Newbie Island?" she asked.

"Do a search on the word 'island,'" Lovie suggested. But they got nowhere.

"Do a search on the word 'Inquiry,'" Father Paul suggested. Again they got nowhere.

Every idea failed until Raquel came up with a different approach. "Accountants require financial transactions to leave a money trail so audits can be performed on all transactions."

"What does that mean?" Father Paul asked.

"Well, this island has got to be expensive to operate. They need food...fresh water...electricity...and all kinds of supplies. So if we find the money trail, it will lead us to the Island."

"Way to go, Sherlock," said Lovie, complimenting Raquel with a pat on the shoulder.

"Click on the Euro," said Father Paul pointing to its symbol.

One click and the screen changed to something that looked like a big, impregnable vault, "This looks like the accounting department to me," said Lovie.

"My, my, challenges and more challenges," said Raquel. "Now let's think like a computer programmer...we are looking for land acquisitions. So if we search for *assets\ real estate\island* maybe we'll get lucky and get a hit."

She pressed *enter* and up popped a strange code prompt. It was the image of a desperate woman alleged to be a witch tied to stake with kindling around its base. The prompt flashed as it waited for a response.

Father Paul almost choked. He couldn't put his grapes down fast enough, "Oh, God have mercy! That's a reference to the Inquisition."

Raquel said, "It's also the Inquiry's firewall. If it accepts the monsignor's password, we will be inside their network. Do you wanna try it?"

"We've never been this close," said Lovie.

"We're getting close enough to breathe down their necks," said Raquel, excited by the adventure.

"We're the heroines, aren't we? " Lovie asked.

Raquel nodded, *Yeah.*

"Wait, wait...ladies, please don't do anything until we give this some thought. I don't know the relationship between the Inquiry and the Inquisition, but I have a bad feeling about this."

"Go ahead, Father. We're listening," said Raquel.

"The Church's most disgraceful years were when the Inquisition was in power. This means we will be probing into the affairs of some very nasty people."

"That's what Seth said before they killed him," said Lovie.

Father Paul pointed his finger at the flashing icon. "That's more than a prompt it's also a warning. Do you heroines still want to continue?" he asked."

"They've already tried to kill you and we crossed the point of no return when we received the captain's logbook," said Lovie."

"Do you want to join the heroines?" Raquel asked Father Paul.

"I'll do whatever is necessary to protect the children," said Father Paul.

"So let's enter the site and get out of there as soon as we can," said Raquel.

"The monsignor probably uses the same password for both networks. So let's give it a try," said Lovie.

Raquel replied by entering the monsignor's password. Then, lo and behold, a majestic-looking book called *The Chronicle* appeared on the monitor!

20. The Chronicle

Raquel swiped her hand over the PenSet, causing the book to open and reveal the first chapter: "In the Beginning."

"This is a wide book," said Lovie. "Is there a way you can display two pages at a time without making the type too small to read?" she asked.

"How's does that look?" Raquel asked after increasing the book's magnification to use the entire screen.

Father Paul, who was sitting to the left of Raquel, put on his eyeglasses. "I can read the page in front of me, but the page in front of Lovie is too far away."

"—Father Paul, if you see something significant in front of you, read it aloud and I will do the same, if it's okay with you," said Lovie.

Father Paul agreed and began to read from the *Chronicle*. "For starters, it says that in the late seventeen hundreds, the church thought that gays were the work of the devil. Hey, listen to this malarkey. During the height of the Inquisition, a group of priests became disgusted with the public behavior of several gay villagers."

Lovie said, "It says here that about that same time, a group of midwives horrified by the birth of a child with both male and female genitals confessed their fears to the same priests." She continued with the next paragraph. "The priests put two and two together and blamed it on Satan. Then the priests took their concerns to the Inquisition, which helped them form a hush-hush affiliate called 'the Inquiry.'"

"Well…now we are getting somewhere," said Raquel, who turned to the next set of pages.

Father Paul eagerly read the text in front of him. "Known only to a few, the Inquiry had one objective: to confront Satan whenever they thought he forced people to act different from what was expected of normal men and women."

"The Inquiry launched a moral cleanup campaign by fighting Satan the only way they knew how," Raquel read.

"…and how was that?" Lovie asked.

"They arrested anyone whose behavior deviated from the norm," Father Paul replied.

"Like, what's normal?" Lovie asked.

"You're asking the wrong person," said Raquel.

"The courts sent most of the accused to the dungeons where they became hospital patients of the Inquiry's medical staff," said Lovie, reading from the page in front of her until Raquel turned it.

Father Paul said, "It says here…this kind of madness continued for a hundred years or so."

Then a new kind of insanity came across the English Channel in the form of a book called *The Origin of Species*," said Lovie. "Oh…I see what's going on. Click the hyperlink right there…where it mentions Charles Darwin."

21. The Pluralists

The heroes had no idea what to expect as they watched the pages of the *Chronicle* fast-forward to the height of the turmoil caused by Darwin's 1859 publication. When the pages stopped turning, Father Paul used his fingers to make sure he read the text correctly.

"As expected, most religious scholars considered Darwin's theory to be an affront to God...and some still do."

"According to this, his work stirred up quite a disagreement within the Inquiry," said Lovie.

"Most churches denounced Darwin's findings as pure blasphemy." Said Raquel, "...and I don't blame them."

"Yet, right here, it says a surprising number of churches embraced Darwin's theories describing evolution as God's handiwork," said Lovie.

"Apparently, several important members of the Inquiry agreed with those churches," said Father Paul. "It says that several Inquiry physicians experienced a paradigm shift."

"...a pair of what?" Raquel asked.

"A paradigm shift," said Father Paul. "...you know, click-up—a light comes on inside your head and reveals a new avenue of thought."

"You must be right, because the Inquiry doctors issued a decree that shook the foundation of the Church," said Lovie. "First they reaffirmed their belief in God Almighty and in the same breath they declared Darwin's theories of evolution to be correct.

"Look here," Raquel said excitedly, "a group of people who embraced Creationism *and* Darwin's theory formed a religious coalition known as the Pluralists."

"Mmm-hmm...so that's how they got started," said Father Paul. "I know some Pluralists. They are God-fearing people, and my, oh my, have their numbers grown!"

"I've read about them," said Raquel. "Their followers come from the world's largest religions. Some are Buddhists, some are Christians plus there are Jews and Muslims amongst them. The important thing is that they all work together."

"Tolerance of their religious differences and interfaith dialogue has made the Pluralists a counterweight to sinister organization like the Inquiry," said Father Paul.

"Ah...here's a slap in the face if I've ever seen one. The Inquiry doctors dumped the Inquiry and joined the Pluralists," said Lovie.

"Right here, it says...the doctors' exodus left an awful taste in the Inquiry's mouths and they swore to get even," said Father Paul.

"So the physicians went one way and the Inquiry went another. Which path do we follow?" Raquel asked.

"Can we search for the word 'Island' in this book and see if we have better luck?" Lovie asked.

"Let's find out," said Raquel, who tried it.

"Well, there it is," she said. "The Pluralists purchased an island from the Philippine government in 1946, which is about the time World War II came to an end."

"Look at this footnote," said Lovie. "They bought the island shortly after receiving large contributions from several major religions."

"It's great to see our feuding religions agree on something, but cohabiting on an island is a stretch for me," said Father Paul.

"I wonder what could draw such diverse religions to an island," said Lovie.

"The island is probably in the Pacific Ocean, but where?" Father Paul asked.

"I've got a hunch. Search to see if the term 'DNA' is mentioned anywhere," said Lovie.

Raquel complied. A moment later she yelled, "We got a hit! In 1995, the Pluralists funded a genetic engineering program," she said.

Father Paul said, "My friend, Father Thomas, the priest who disappeared, told me about that. But according to him, the project blew up in their faces. The Pluralists hired that geneticist...oh, what's his name. Scorn something or other...ah...Dr. Liborio Scorn, that's it. Anyway they hired him to lead the team, but Scorn had a mission of his own."

"Like what?" Rachael asked.

"Apparently, Dr. Scorn did not like the direction Mother Nature was taking humanity. So he decided to steer us toward a more perfect human species."

Lovie read a paragraph from the *Chronicle*. "According to this, Dr. Scorn had a particular interest in people carrying what he described as the 'Gay DNA Strand.'"

"What's that?" asked Father Paul. But everyone had the same question.

Raquel read a paragraph that said, "Scorn needed volunteers, so the Inquiry offered a bounty for people who carried the gay DNA strand."

"Oh boy, will you look at the number of hospitals that jumped on Scorn's bandwagon after the reward was announced?" said Father Paul.

Raquel said, "Hospitals in major cities across the globe collaborated with the Inquiry."

"My, my, what a shame, to see hospitals behave like hoodlums," said Father Paul.

"I think we should upload this", said Raquel.

Lovie agreed. "I bet this is when the Cyclops started transporting gay slaves."

"This makes me sick," said Raquel, who fast-forwarded past several pages to skip the sordid details.

"Whoa…hold on a minute," said Lovie, urging Raquel to slow down. "Did y'all see that?" she asked, spinning her index finger counterclockwise. "Go back a few pages. That's it…keep going…back a little farther…stop right there," she said.

Raquel read, "The Pluralists fired Dr. Scorn in 2013 and replaced him with a scientist they could trust, but they don't mention a name."

"Wow…will you take a look at this?" said Lovie. "Apparently, in 2015, their new science team leader got impatient for answers and started an accelerated breeding program."

"That was forty years ago. We need to find out more about that," said Father Paul. "No telling what's happened since then."

"We'll have to switch to the Pluralists' website to find out," said Raquel. "If anyone objects, this is the time to voice it," she said, waiting with her finger poised to press the *enter* button. When no one did, she switched to the Pluralists' site.

The Pluralists' visitors' homepage appeared, offering members the opportunity to sign in with a password. "Here we go again," said Raquel as she entered the monsignor's password.

The member's homepage showed that the monsignor had several unread e-mails. But the advertisement offering an opportunity to 'Invest in Newbie Island' grabbed everyone's attention.

"Will someone buy some popcorn? Because we are going to the movies," said Raquel. Then she clicked an icon resembling an old-style Hollywood movie camera.

22. The Science Team

The Pluralists' webpage used excellent sound effects. Our heroes heard the crackly noise of celluloid film moving through an antique movie projector while they waited for someone to appear on the screen.

Finally, Abu Jubar, a popular political figure from Calcutta appeared and he was all smiles and eager for them to watch the video. "Welcome! I bid *wel-come to our good friends and new investors. The Pluralists' science team is well on the* way to understanding humanity's destination. Although our pursuit of knowledge is expensive, we expect a handsome payback. That's why we are seeking investors like you to share the expenses…and reap the rewards.

"You probably have read our proposal and earnings forecasts. Now we have something special for you. In a moment, the leaders of our science team will give you a glimpse into our future. So get comfortable. They will join us in a moment."

Abu waved good-bye and the screen blacked out for a moment. When the video resumed, there were three scientists standing in a large mezzanine welcoming the viewers to Newbie Island. They waved vigorously for the video camera to come in for a closer look. All of the scientists wore white lab coats; however, judging by his white collar, the one standing in the middle was a priest as well.

Father Paul recognized him as Monsignor Mancini and he gasped at the thought of what this implied. "I hope my eyes are deceiving me," he murmured.

"There's nothing wrong with your eyes, Father," said Raquel.

"I warned you guys not to trust him," said Lovie.

"Well, I'll be," said Father Paul. "The monsignor has his hands in all sorts of activities. It makes you wonder what's going on in the Church."

"He's a Catholic, he's a Pluralist, and he's leading a scientific mission that may conflict with his beliefs," said Raquel.

"I bet 'Mr. Everything' is a member of the Inquiry also. I like him less every time I see him," said Lovie.

"I can't imagine him in the Inquiry. But what I'm seeing right now makes me wonder," said Father Paul.

"But what if he participated in the atrocious acts mentioned in the logbook?" Lovie asked.

"Have either of you ladies read the logbook from beginning to end?" Father Paul asked.

"It's too painful to read," said Raquel.

"Then we don't know if he was involved," said Father Paul.

"To be safe, we will consider him guilty until evidence proves him innocent. I don't like it, but we have to watch our backs," said Lovie.

"...agreed," said Raquel.

"I feel like a deflated tire!" said Father Paul. "I've always admired the monsignor. We didn't agree on some issues, but I considered him a credit to the Church, but here lately, he—"

"Shh..." said Lovie, like a librarian quieting the kids...he's getting ready to say something."

"Welcome to Newbie Island. I'm Monsignor Mancini. I'm a priest, a doctor of biochemistry, and the science team's leader." He nodded briefly and turned to the female scientist standing next to him. "My colleague to the right is Doctor Susan Minh, the leader of our genetic engineering team…and to my left is my dearest friend, Doctor Kruel, whose specialty is molecular evolution. We are here to do a little boasting about our achievements." He grasped the hands of his colleagues and they bowed together.

It was an impressive preamble—three world-renowned scientists collaborating—and they could not have selected a better backdrop to discuss where evolution was taking humanity. It was no coincidence that they were standing in front of a floor-to-ceiling mural showing several prehistoric hominids walking a path called 'Evolution Road'. The figures walked single-file, in chronological order, each representing a link in the human chain of evolution. The bronze tags beneath the last three identified them as the Neanderthal, Cro-Magnon man and Modern Man followed by a question mark.

Msgr. Mancini pointed at the wall, "…this is a magnificent mural even though the most bewildering feature is the question mark. Can anyone tell us what it means?" he asked.

The monsignor bent forward and cupped a hand behind his ear to give those watching the video the impression he was straining to hear a faint voice from someone in the audience.

"Oh…yes, now I can hear you. So, you want to know the meaning yourself? Well maybe one of my esteem colleagues can explain the meaning of the question mark," he said bowing to the female on the team.

Dr. Susan Minh was Hong Kong University's most famous alumni. She graduated summa cum laude before she was twenty. Then for the next several decades she earned accolades for her pioneering work in genetics. "Actually, I'm reluctant to express an opinion because quite often the 'truth' is a hard pill for some people to swallow," she said.

EVOLUTION ROAD
NEANDERTHAL
CRO-MAGNON
MODERN MAN

"She's right," said the monsignor. "It's always been difficult for scientist to reveal a 'truth' that contradicts well-established thinking...here's an example. *There was a time when* everyone believed the world was flat. Therefore you can imagine the public's outcry to a scientist saying the world is round."

"We call that a hard-sell," said Dr. Minh.

"Maybe Dr. Kruel can put us on the right track. Where is evolution taking us?" the monsignor asked.

Dr. Kruel was a brilliant Russian scientist known for two things. His experiments which often ventured very close to the boundaries of medical ethics and he was also known for having a marvelous sense of humor that relaxed people when tensions were high. His comical antics went well with his unusual appearance. His mustache, eyebrows and the hair on his head was snowy-white. Years ago he explained that his hair turned white all at once when a frightening experiment went completely wrong. Finally, he spoke, "Can you imagine a brave scientist sticking his neck out trying to convince a crowd of people that Earth and everything on it was rotating on its axis and revolving around the Sun, even though the crowd he was addressing did not see or feel any evidence of movement?"

"Wow," said the monsignor smiling at the predicament. "Back then a scientist could be sentenced to the loony pen for telling the truth."

"Now, here's an even tougher truth to sell," said Dr. Minh. "Right now, every one of us is experiencing subtle changes inside our bodies. The changes occur at the genetic level. We can't feel it anymore than we can feel the Earth's rotation...but we are evolving. We are changing from what we are—-to whatever the question mark represents."

"Which brings us to why we're here on Newbie Island," said the monsignor. Several years ago, the Pluralists' Church realized that humankind was morphing into something different. *But what...they wondered.* No one knew, so they asked us to find out."

"To be safe, we asked the world's scientific community first," said Dr. Kruel, "but they could only speculate."

"Dissatisfied with the status quo, we decided to find out for ourselves," said Dr. Minh.

"Which was a lot easier said than done," said the monsignor, wiping his brow as he thought about the obstacles they had overcome.

Dr. Kruel said, "Maybe this will give you an idea of the project's complexities. For starters, we needed hundreds of volunteers to live and breed in an intelligent, controlled environment...for the rest of their lives."

"But first, we had to design and build the controlled environment, which cost so much money that we had to go back to the Pluralists' Church for funding," said the monsignor.

"The Pluralists did not flinch at the projected cost. No sirree, not them. They went out and hired the best engineering minds available and got the job done," said Dr. Kruel.

"To call our controlled environment a 'laboratory' would be an injustice, because a community of people lived inside it," said Dr. Minh. "I have a picture of it stored in my PenSet. Here, I will show it to you."

23. Laboratory City

Dr. Minh displayed a picture of an enormous structure. It was a three-story facility that spanned the island from one end to the other. A huge, transparent dome supported by rafters covered the entire building—which in itself was not a small feat, since it occupied seventy percent of Newbie Island.

"Can you can see why we called it Lab City?" asked Dr. Kruel. "Our volunteers lived inside, so consequently we had to make it very comfortable. There were schools, and homes with grassy lawns, flowers, and plenty of trees."

The monsignor said, "Quantum bit computers regulated the environment to insure that the air was fresh and the temperature and humidity were ideal for comfort."

"It had to be comfortable, because the volunteers were not allowed to come out, and of course we were barred from going inside," said Dr. Kruel.

"Lab City was built with features that allowed us to study the volunteers at the genetic level," said Dr. Minh. "As a result, we were able to observe and record any and all changes over an extended period of time."

"Lab City was the world within which our volunteers and their offspring lived, bred, and eventually died," said Dr. Kruel.

"We solicited volunteers from across the globe," said Dr. Minh.

"It was important for us to have volunteers of all ages and races. So in some cases, we accepted entire families," said Dr. Kruel.

"The compensation was so attractive that we got more volunteers than expected," said the monsignor. "But…that turned out to be a lucky break because it gave us the opportunity to be selective."

"We tripled our genetic engineering staff," said Dr. Minh. "And we installed a network that gave us access to related projects being worked on by scientists elsewhere."

Dr. Kruel said, "We tested the volunteers before we brought them to the Island."

"We tested their physical stamina first and those who passed went on to take a rigorous psychological and intelligence test," said Dr. Minh.

Dr. Kruel said, "Seventy percent of the volunteers failed one test or the other. The volunteers who passed both tests were brought to the Island and split into two groups."

"The first group came from bloodlines with a history of breeding only straights," said Dr. Minh. "And those who came from bloodlines with a history of breeding gays and straights went into the second group."

Like any ordinary people the volunteers experienced both stormy and pleasant days as they journeyed from birth to death—with one big exception: our volunteers lived their lives at an accelerated rate," said the monsignor.

"In retrospect, would you say that figuring out how to speed up evolution was our most difficult obstacle?" Dr. Kruel asked Dr. Minh.

"Absolutely, the truth is that Mother Nature's rate of evolution is too slow," she replied. "If we were going to make progress within our *own* lifetimes, we knew we had to speed things up."

"We had two options," said Dr. Kruel, using comical gestures to make his point. "One of them required us to figure out how to leap forward in time, study the next link in our chain, and then somehow get back here, in one piece."

"That idea lasted less than a minute," said the monsignor, smiling.

"Fortunately, a pioneering pharmaceutical company came up with the Brutus Inhaler, an aerosol that sped up the life cycle of human genes," said Dr. Kruel.

"This drug gave us precisely what we needed. Instead of us traveling as Dr. Kruel mentioned, we watched the genes of our volunteers travel Evolution Road step-by-step. We observed the minutest genetic changes as well as the physical changes brought about as a result," said the monsignor.

"This is Dr. Minh's field...so, Doctor, maybe you can help us understand this?" Dr. Kruel asked.

"Under normal circumstances, our genes lead very active lives," she said. "However, after the volunteers inhaled the Brutus aerosol, their genes became even more active. What they had previously accomplished in an hour now only required a fraction of an hour. It was astonishing. The drug sped up the life-cycle of our original volunteers. They went to school, got married, raised families, and passed away, but they did so at a rate considerably faster than yours and mine."

The monsignor seemed pleased with his team as he said, "The volunteers' children completed the same cycle at an even faster rate...and so it was for every succeeding generation. The Brutus aerosol gave us the ability to control the speed of evolution inside Lab City. It entered the laboratory via the heating and air conditioning ducts but of course the volunteers were aware of this"

"We got to the point where a hundred generations were born and lived their lifetimes within a single day," said Dr. Minh.

Dr. Kruel said, "The Pluralists wrote and enforced a Supreme Doctrine that prevented us from altering the course of evolution...and as God is our witness, we obeyed that law."

"It was incredible," said Dr. Minh. "We witnessed the life-cycle of many more human generations than we ever expected."

"Following the genes as they traveled Evolution Road was fascinating to watch," said the monsignor.

Dr. Minh's excitement was evident as she spoke. "I always knew our genes had intelligence, communicated with each other, and made decisions, but...I...ah, I never imagined I would be a witness. So I was honored to be a part of all this."

"We watched and recorded the evolution of the volunteers and their genes. And what we've learned is mind-boggling," said Dr. Kruel.

"Let me put it this way," said Dr. Minh. "We witnessed a thousand years of human evolution within thirty calendar months. Can you imagine that?"

Monsignor Mancini looked into the camera and asked the potential investors a poignant question. "We got a glimpse of man's future and we recorded it.

Do you have any idea how valuable our laboratory recordings are? 'Priceless' would be an understatement when you consider the valuable education they offer to the physicians and scientists of the world."

Usually Dr. Kruel had a happy face, but for some reason he now appeared sullen and worried as if something unpleasant had crossed his mind. "We also learned a valuable, but unexpected, lesson as well," he said. "Our speeded-up version of evolution proceeded according to plan. That is... until the genes of our volunteers encountered a hazard on Evolution Road."

Again, Dr. Minh used her PenSet to display a road map showing the location of the road hazard, "It took us six months to travel two hundred years up Evolution Road but when we got right about here...we encountered a serious challenge."

"Have you ever heard of a genetic merger?" the monsignor asked.

"The merging of species is an integral part of evolution," said Dr. Minh. "A classic case in point would be the genetic merger that occurred when the dinosaurs known as velociraptors became birds of prey."

"Now, that was a wedding and a weeding-out process, if there ever was one," said the Monsignor. "The genes of numerous dinosaurs became extinct because they could not adapt to the changes in their environment."

Dr. Kruel said, "A better-known genetic merger occurred when the genes from various ape species merged and became humankind. In this case, an enormous number of ape genes became extinct because they could not adapt to the changes."

The monsignor said, "My colleagues are trying to tell you that we humans have a very unpleasant merger in our future."

"I'll explain," said Dr. Minh. "When the genes of our volunteers reached this place on Evolution Road, the male and female genes were required to merge in order to continue evolving."

"For the lack of a better name, we dubbed it the Gender Merger," said Dr. Kruel.

The monsignor said, "It started when the genes of our volunteers converged on a narrow, tunnel-like passage. At first, gaining entrance was competitive, but then it got messy—and before long, it became deadly."

"It's not pretty. But we created an animation to demonstrate what happened," said Dr. Minh. She pressed the *play* button on her PenSet and a video began.

The 'genes' unique to the male and female genders, shown in blue and pink respectfully, did something similar to converging in the narrow entrance. It was as if the genes rushed foolhardily toward the opening when they should have been negotiating rights-of-ways. They took an awful beating by crashing into each other at high speeds. The effects was somewhat like the genes had burst and scattered into pieces, starting a chain reaction of new collisions. It was awful!" Doctor Minh pointed to a pile of wrecked 'genes.' "The chaos you see at the entrance of the tunnel continues all the way through."

Dr. Kruel said, "I'm glad to say that the genes of twenty percent of the children born to our volunteers came out the other end of the tunnel unharmed."

"...that's a diplomat's way of saying that the Gender Merger may very well become our apocalypse. We predict that eighty percent of the human race will not survive the ordeal unless something drastic is done," said the monsignor.

"It's unpleasant but it's something we must face," said Dr Minh.

"If you are wondering why the twenty-percent survived...I will tell you... they had something in common, said the monsignor.

"They all carried a strand of DNA common to gays," said Dr. Minh.

"It comes down to this," said Dr. Kruel. "Mother Nature in her infinite wisdom blessed some people with the Gay Strand. If she had not, none of our volunteers would have survived the Gender Merger.

"As it stands now, most human bloodlines don't have it and as a result they are incapable of completing the Gender Merger," said Dr. Minh.

"Unless...our good friend, Mancini, finishes that Serum he's been working on for twenty years," said Dr. Kruel, giving the monsignor encouragement by patting him on the shoulder.

"We'll have more to say about the Serum after we complete the next round of testing," said the monsignor. "In the meantime, I believe our investors would like to meet the offsprings who made it through the tunnel."

Then he took a remote control from his lab coat pocket and aimed it at the mural. *Click-up* was the sound the heroes heard before a section of the wall opened and revealed the next link in the human chain of evolution.

24. The Newbies

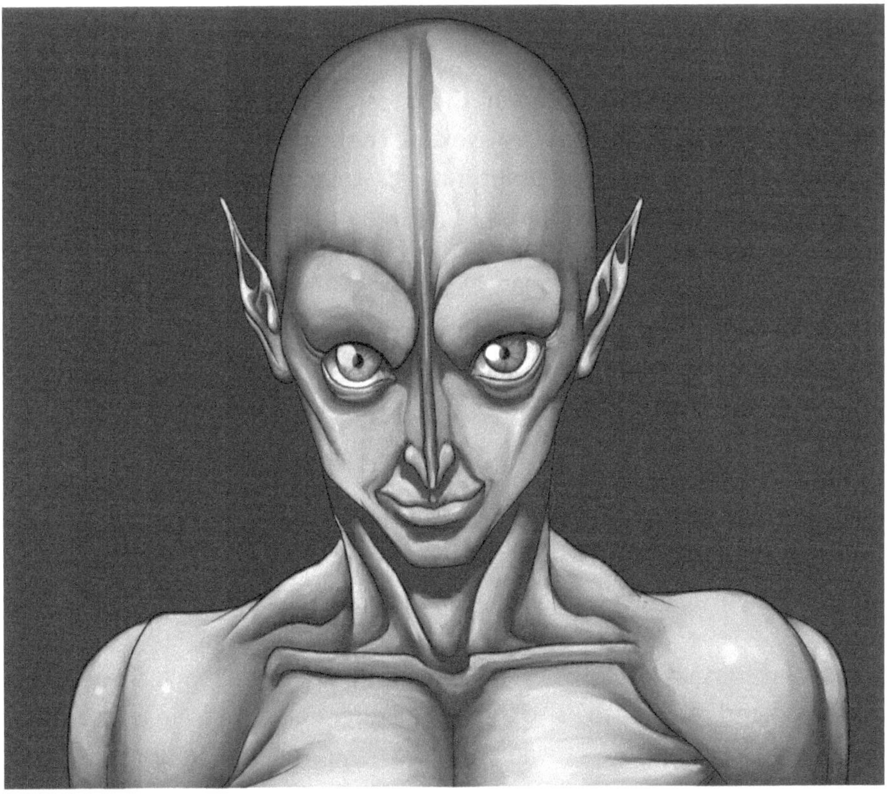

A section of the mural wall disappeared and in its place a Newbie appeared on a raised platform, alive and buck naked. The sight of the nude creature embarrassed Father Paul, who instinctively closed his eyes but after a moment he peeped through his fingers and studied it. The women's reaction was different. The instant they saw it, they looked at each other because they realized the person who had called himself Saylor was a Newbie.

The creature appeared to be about six feet four inches tall and it had a hard-edge that ran a course from the center of its skull to the tip of its nose. The body was muscular, the skin looked smooth and there was no hair anywhere.

The Newbie's genitals looked feminine. The lips and nose were keen. The eye sockets were large and the eyes were green. The creature was expressionless but the eyes were piercing, as if probing the depths of the of the hero's hearts and souls in search of their innermost secrets. They could tell that it was thinking. A slight smirk gave the impression that it knew something they ought to know.

Dr. Kruel volunteered to answer the most frequently asked questions. "We call them Newbies. They are asexual. There are no males. There are no females...just Newbies. Like other humans, they have distinguishable features that give them individuality—facial features, character, size, color—about the same as you and I."

"They are born pregnant. Their embryos gestate for fifteen earth years. At year thirteen, they mature and breasts for suckling their young appear. Two years later, the child is born," said Dr. Minh.

Father Paul said, "Oh, my-gosh. Humans are going to become a single-sex something or other that obviously does not have sex with others."

"They've taken all the fun out of making babies," said Lovie.

Msgr. Mancini said, "Newbies are intelligent beings. We teach them about ourselves, and they teach us to think in all directions. We have learned that they are pacifists opposed to violence as a means of settling disputes."

Dr. Kruel said, "To them, thought is a vessel for exploring possibilities. Tests show their IQs to be about...ah...roughly one-and-a-half times ours. Would you agree, Doctors?" he asked. The other scientists nodded.

"Their senses are rather keen," said Dr. Minh. "Their nerve cells interface with those of other Newbies, creating a neural network that operates at *extremely low frequencies* known as ELF."

"They cannot grow hair because ELF creates electrolysis—an electric current that passes through the roots of their hair and removes it," said Dr. Kruel.

The monsignor said, "ELF is their equivalent of a wireless telephone network. Everyone enjoys connectivity to each other. As a result, Newbies rely less on the spoken word and more upon thought transference."

"They can take a poll or vote on a controversial issue in a few seconds," said Dr. Kruel.

"Humans cannot hear ELF transmission, but other creatures seem to hear it quite well," said Dr. Minh.

"In other words, the Newbies can talk with us and all the other creatures that inhabit our planet," said the monsignor.

Doctor Minh said, "Their diets are drastically different from ours. They considered eating the flesh of other creatures primitive and distasteful. Vegetables are acceptable, but they prefer a crustacean called krill."

Suddenly a red icon flashed on the screen, signaling a message from the Inquiry. It startled Raquel who pushed her chair back from the table as if the message were a viper poised to strike a death blow, "Whoa...why and how could the Inquiry send us a message?"

The flashing message was a general broadcast notifying users that the Inquiry's network was experiencing high traffic and that several users were waiting to log on.

"How can this be? We left their network, didn't we? "Lovie asked.

"I thought so but…maybe I didn't sign off before logging onto the Pluralists' network," said Raquel.

"Uh oh, it's time to go. Let's get out of here," said Father Paul.

"I second the motion," said Lovie.

"I've uploaded the Lab City files and I'm about finished uploading the image of that Newbie to your cloud account. There we go…now I'm ready to high-tail it out of here," said Raquel, completely unaware that their activities were under human surveillance.

25. The Inquiry's Computer Center

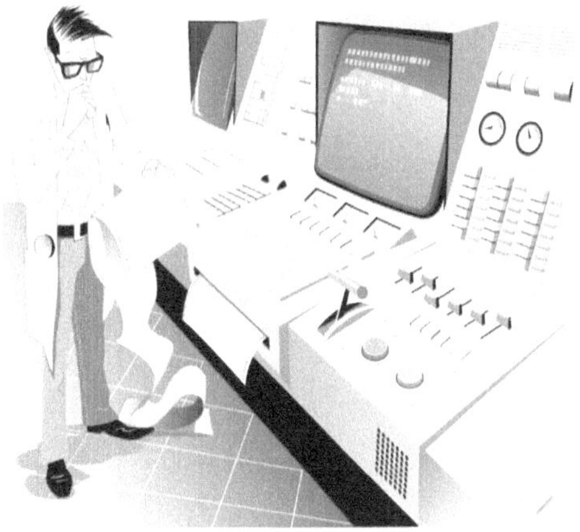

The Inquiry's mainframe computer system ran 24-7 and was under the constant watch of an operator.

Tonight, the system was unusually busy. Raquel's uploads had caught the operator's attention because only a few users had such authorization. When he checked and the computer identified the user as Monsignor Mancini, the young man gave no more attention to the activity.

Although the Inquiry required an operator to be in the computer room at all times, trips to the restroom were the exception, and the operator had to go. The young man rushed to the men's room and hurriedly used the urinal. He noticed drips on his trousers when he zipped up his pants on the way to the face bowl.

A toilet flushed while he washed his hands in the sink. Instinctively, he looked into the mirror and there was Msgr. Mancini coming out of a stall, "Pardon me, Father did you finish the upload to your Cloud account?"

"What Cloud account…? "The monsignor asked, absolutely dumbfounded by the question and by the young man's quick exit from the men's room.

26. The Rescue

The 'heroes' were preparing to leave the church's conference room when Saylor burst through the door. "Are you all right?" he yelled, as he looked from wall to wall and around the room, poised and ready to react to the slightest sign of danger.

His intrusion most certainly shocked the heroes. They looked like cartoon characters, their faces frozen while their minds tried to recover from seeing Saylor-the-Newbie burst through the door. It was intimidating! They weren't sure if they should run like hell or accept his offer of protection. Then a courageous Lovie stepped forward and took control of the situation.

"What's going on? Where are the guards?" she asked, while trying to look over Saylor's shoulder and into the hallway to see for herself.

"Your guards are safe, but you are not," said Saylor accusingly.

"We can take care of ourselves," said Raquel.

Saylor ignored her naiveté and said, "The three of you hacked the Inquiry's computers and they're very angry. A few hours ago, their thugs paid a visit to your condominiums and burned them to the ground.

"What?" Lovie exclaimed in disbelief.

"Right now they are moments away. If you want to live...come with me," was all he said before he ran out the back door.

"As you can see, Saylor is a person of few words. But I believe him," said Lovie. "Come on Raquel. Get your camera ready, we're going to interview a Newbie." Then she rushed out the back door after Saylor.

The church's hilltop position provided a remarkable view of the ocean for miles around, but there was no time to enjoy it now. The heroes were too busy trying to keep up with Saylor.

"Be careful," he said, as he led them down a steep and winding dirt path dug out from the hillside. The path was so steep it forced them to run faster than they preferred. Running into each other was unavoidable—sometimes it was the only way to stop in time to make the next curve. No one got hurt, but it didn't take long to reach the beach where Saylor had a speedboat ready to go. As soon as everyone boarded, he sped from the shore, steering due north along the coast.

The heroes had a lot on their minds. They were trying to catch their breaths from the downhill slalom but their minds were on Saylor and the decision they made to place their lives in his care. All eyes were on him, but unfortunately since his back was turned, they couldn't see his face. To make matters worse, he wore a knitted cap that covered his ears. There they were, second-guessing themselves and wondering if placing their lives in the hands of a Newbie had been the smart thing to do.

"Are we in the frying pan or the fire?" Father Paul asked.

"We're about to find out. He's slowing down," said Raquel.

They were about two miles up the coast when Saylor steered the boat into a small cove. The night was quiet and a slight breeze caressed them. "I'm stopping here because our pursuers cannot see us from the church's lofty position. Now, where would you like to be dropped off?" he asked.

Lovie ignored his question to ask one of her own. She gave Raquel the signal to start recording while she interrogated him. "What did you mean when you described the guards as 'safe'? What did you do to them?" She asked.

Raquel had the PenSet's camera aimed and recording and Saylor seemed amused by their antics.

"You're wearing that cap and those big sunglasses because you're one of them, aren't you? You are a Newbie I bet you are!" said Lovie gazing at Saylor's as if a flinch or gesture would let her know when he lied.

"That I am," he calmly replied as he removed his sunglasses. "...as for your guards...they are unharmed, but they will sleep for a while."

Raquel said, "Under normal circumstances this would sound a bit weird, but...may we see your ears?"

Saylor grinned, removed his cap, and revealed his pointed, feline ears. "Is this proof enough for you?" he said, wiggling his ears at the camera.

"My, gosh I know you're called Newbies," said Father Paul, "but what in heaven's name *are* you folks?"

"We're human beings, like you."

"Not, hardly," said Raquel.

"Then look at it this way. Earth is the home of men, women, and Newbies."

"Do you feel like a man or a woman?" Lovie asked.

"I have no way of knowing," said Saylor.

"...Probably a man," Raquel ventured to guess.

"Probably a bit of both," Saylor countered.

"Tell us about your people," said Father Paul.

"I presume that you have as many questions about us as we have about you, but this is not the right time," said Saylor. "The Inquiry has placed a bounty on your heads, so I hope hacking their network is worth the trouble you've caused."

"It was worth the risk," said Lovie, who received affirmative nods from Father Paul and Raquel.

"Where are *you* going?" Father Paul asked.

"My ship returns to the Island tomorrow."

"May we go with you? Lovie asked. "The Inquiry will be waiting for us onshore, and I need to interview your people."

"That idea might be possible…if we can sneak aboard the freighter before the crew returns," said Saylor, who resumed the journey north along the coast.

"Where is the crew?" Father Paul asked.

"Most of them, including yours truly, are on shore leave," said Saylor. "There are only two guards on duty and they're about to get phone calls that will draw them away from where we need to board the ship."

"Ahoy, there's our transportation straight ahead," said Saylor, pointing to a ship aglow in the moonlight. The vessel had the long, sleek look of a luxury yacht with elegance glimmering from its white porcelain exterior.

"Wow, I've seen those kinds of ships in magazines," exclaimed Father Paul.

Saylor said, "You are looking at the fastest freighter on the planet. Cavitating propellers allow it to skim the ocean's surface at speeds in excess of eighty knots."

"The front of the ship look like the mouth of a sea monster," said Lovie.

"That's because it's a catamaran. But that section over there resembles something a hovercraft would use," said Raquel, pointing to a section of the freighter that had skirting.

"You know your ships," said Saylor. "This ship has the air-cushioned ride of a hovercraft and the twin hulls of a catamaran."

The communication hardware reminded Raquel of her father's yacht. "It looks like Santa left a collection of big-boy toys on the rooftop," she said.

"I think those toys were too big for Santa to take them down the chimney," said Father Paul.

"What are those?" Lovie asked pointing to the colorful crates stacked on the aft deck.

"We're delivering those crates to the Island in the morning," said Saylor, turning off the speedboat's motors before those onboard the freighter could hear them. Then he used trawlers to move them quietly toward the freighter's loading dock.

"Now back to the phone calls we have to make," said Saylor. "Father Paul, call this number and you will cause a guard to go to the machine room. Raquel, I want you to wait thirty seconds. Then call this number, which will draw the other guard to the wheel room."

"What do we say when they answer?" Father Paul asked.

"Delay them as much as you can. You will be sneaking onboard while you are talking with them. Start calling now and follow my lead," said Saylor.

The heroes could hear a phone ringing when their speedboat pulled up to the freighter's loading dock.

Saylor whispered, "Go up those steps quietly and wait for me. I'll be with you as soon as I put this boat in the garage."

Unbeknown to the heroes, a third crew member, Billy Ray saw the whole charade. He watched them sneak onboard and he recognized the heroes from a mobile phone commercial offering a handsome reward for them, dead or alive.

27. The Freighter

The ship's engine room was a maze of machinery in the hands of obsessive engineers who treated every electrical and mechanical component as if it were a blood relative. The engineers came back from shore leave eager to start the engines. In fact, they were so busy they did not notice the meeting occurring in a dark area around the corner. Three members of the deck crew—Billy Ray, Joe, and Shep—were having a discussion about Saylor, the stowaways, and the reward money.

"I tell ya, they are the same people," said Billy Ray, holding up his mobile device so his cousins could see the picture of the heroes.

"See for yo 'self," he said. "This here's the picture that came to me from the Inquiry…and I swear that *this* is the picture I took when they snuck aboard."

"There ain't no need of you swearing Billy Ray, 'cause your mouth ain't no prayer book. Y'all can count me out," said Joe, shaking his head and waving off the idea of capturing the heroes.

"I'm convinced," said Shep.

"This here is easy money. They are wanted dead or alive and we've got 'em trapped on this here ship," said Billy Ray.

"What about Saylor?" Shep asked.

"I need the money. So we'll kill anyone who gets in the way," said Billy Ray.

"I don't care whatcha say. I'm not gonna risk my job on a wild hunch," said Joe who emphasized his point by pushing his fingertips into Billy Ray's chest.

"No one's begging you, Joe. It's strictly yo' choice," said Shep. "But if Billy Ray and I earn some life-changing money, we don't want to hear you boo-hooing about being left out. Is that clear?"

"How much y'all figure on getting?" Joe asked.

"…a heap, "said Shep.

"Okay, count me in. But you betta' be right," said Joe.

28. Sleeping Quarters

The heroes followed Saylor to one of the ship's cabins—upstairs to the next level and down a corridor—until Saylor stopped in front of a cafeteria midway down the aisle.

"This is the crew's diner. There's food in here for making soups and sandwiches if you're hungry. C'mon, your rooms are farther down this way," said Saylor.

He walked a few more paces and stopped in front of a door. "This is the guest cabin. All of you can sleep in here. It's larger and it has two bedrooms—" Suddenly, the roar of the ship's engines made it difficult to hear Saylor.

The engines sent vibrations crawling up the heroes' legs. Then they heard the screech of ear-piercing whistles. Gradually, they felt the ship moving—slowly, at first. But when the engines began to purr, the ship picked up speed and cruised out to sea.

"I'd better fetch some sandwiches and cold beers before the kitchen close," said Father Paul. "What can I get you, Saylor?"

"Some other time," Saylor replied, and walked away.

29. The Thugs Attack

Billy Ray and Joe grabbed Father Paul when he came out of the cafeteria with a tray of food and drinks. He tried to resist, so they roughed him up a bit before dragging him back to the cabin. When they opened the door, Father Paul was surprised to see that Lovie and Raquel were already captive.

"Do not fear, the Lord is with us," he said. Then Billy Ray struck Father Paul with the butt of his pistol, knocking the priest to the floor. "You'd betta' hope so 'cause we're gonna hang y'all. Then we're gonna throw y'all overboard. Ain't that right?" His cousins agreed.

Shep was having a difficult time tying Lovie to a bedpost by himself. She was on the floor, struggling and trying to kick him, "Lady, it's a good thing yo' hands are tied 'cause I'd have to cut yo' throat the way you're acting," said Shep. He drew his knife and threatened Lovie, "Now, are you gonna behave or not?" He asked, pressing the point of his knife against her throat.

Suddenly, his nose started bleeding and several drops of blood fell on Lovie. "Keep your blood to yourself!" she yelled.

"...whatcha talking about?" Shep asked, but when he saw the blood dripping on her, he dropped his knife and touched his nose to confirm that it was bleeding. Startled, he looked to Billy Ray and Joe for help, but they were bleeding as well—from several places.

Suddenly, the three of them fell to the floor unconscious. The cabin door swung open and there stood Saylor, as inexpressive as ever. The heroes were dumbfounded. They looked from Saylor to the unconscious thugs and back to Saylor. They believed he had something to do with what happen, but how could he have disabled the thugs from the other side of the door?

"Well, this most certainly complicates things," said Saylor. "Stay in the cabin and get some sleep. In the morning I will tell the captain that I found stow-aways. But by that time, we will be too close to the Island for him to take you anywhere else.

"What about the thugs? Father Paul asked.

"I'll get rid of them if you promise to get some sleep. You need the rest. Tomorrow will be a lot more difficult."

The heroes were too exhausted to do anything other than agree. When Saylor departed they flopped down on their comfortable mattresses and slept like logs the rest of the night.

Unbeknown to the heroes Saylor had used ELF to disable the thugs and now he was using it to communicate with the Newbies on the Island, "Our long awaited day of discovery is upon us. The freighter will arrive before noon tomorrow. Subdue the longshoremen, man the docks but keep a low profile until the ship drops anchor."

30. Newbie Island

The cabin phone rang early the next morning. Lovie answered on the speaker so everyone could hear the conversation, "Good morning crew this is Saylor. Are you up yet?"

"Just barely, said Raquel, yawning as she brushed her hair.

We're two hours from the Island, which gives you about an hour to freshen up and meet me topside. I have to go now," said Saylor before hanging up.

"Aye-aye, sir," said Lovie, giving the speaker a naval salute.

An hour later, a bright and sunny morning greeted the heroes on the top deck. The freighter was cruising, making barely a sound as it skimmed the surface of the water at high speeds. The cool breeze carried the refreshing smell of the sea while albatrosses soared high against the clouds and screeched at the new day.

The heroes inhaled deep breaths of air until they saw Saylor waving from the front of the ship. He was in the cargo gear area signaling for them to join him. Their path to him ran between cargo crates that stood eight feet high. "These crates remind me of the office buildings on Wall Street," said Raquel, looking up at them.

"They must weigh tons," said Father Paul, trying to judge the weight of a crate by pushing against it with his frail body.

"Watch your step," said Saylor as the heroes stepped into the cargo gear area.

"Did you speak to the captain about us?" Father Paul asked.

"Oh yes and he's very angry. He plans to talk with you after we dock."

"All I can tell him is the truth," said Father Paul, who raised his right hand as though he was about to testify in court.

"Which reminds me of a question I have for you," said Saylor. "My people love living on the Island, but we need a nation. Do you think the Pluralists will allow us to rule it?" He asked.

"Uh…maybe and maybe not," said Father Paul.

"Since we are the majority on the Island, maybe we should vote on it."

"Democracy doesn't work exactly that way. But I get your drift," said Father Paul.

Saylor pointed at the horizon. "There she blows. That's the Island over there. Isn't it beautiful? Get ready…the captain is slowing down the ship."

Saylor knew his people were in control of the Island, but he didn't want the ship's crew to know it. To them, everything looked normal when the captain steered the freighter toward a pier constructed specifically to handle marine cargo containers.

The longshoremen on the dock gave the impression they were about to unload the cargo. By the time the ships' crew recognized them as Newbies, it was too late. The Newbies transmitted a barrage of ELF signals that knocked all straights on the ship unconscious…including the heroes.

31. The Infirmary

Consciousness returned slowly for Father Paul: waking was like surfacing from the depths of the ocean. He could see people but not recognize them, and he heard sounds but they made no sense. He struggled to open his eyes but closed them quickly when he realized one of the physicians attending him was a Newbie.

"He's awake and pretending to be asleep," said one of the doctors causing Father Paul to darn near faint from the embarrassment.

"Hello, Father Paul...welcome to our hospital. My name is Dr. Anh. I'm your attending physician and Dr. Minh is helping me."

A mortified Father Paul opened his eyes. His throat was too dry to speak. He nodded a greeting instead and immediately began looking for a way to escape. That's when he heard a familiar voice, "Welcome to the Island," said Lovie, lying in the bed next to him. "The Newbies have taken control of the Island."

"You mean a coup?"

"It was the best way to protect the Island from the Inquiry," said Dr. Kruel, who was attending Raquel since she was too groggy to get out of bed.

It took a moment for Father Paul to recognize that the doctors attending them were the scientists they'd seen in the video the day before. It was about then that he heard Lovie yell so loud it frightened everyone on the ward.

"What I don't understand…is why the Newbies haven't put that man in chains!" She yelled at Monsignor Mancini, who was attending an injured member of the ship's crew.

Father Paul noticed the monsignor for the first time and grimaced at the thought of him providing medical care to anyone prior to substantiating his innocence. He sat up, pointed his finger, and screamed, "Get him out of here! You'll never use me for a guinea pig…you swine, you!"

The monsignor grimaced as the priest's sharp words cut at him. "Oh shut up, you old fool, or I will have one of the Newbies stuff your mouth with soiled linen."

Father Paul began to reply, but the monsignor abruptly cut him off. "…and you're free to leave the hospital anytime you want." The offer caught Father Paul by surprise. Baffled, he studied the face of his longtime adversary for telltale signs of dishonesty but there were none whatsoever.

"You're telling the truth, aren't you? I can see it on your face."

Lovie was too woozy to get up, but her voice was working fine. "Dr. Anh, do you know anything about him?" she asked, pointing at the monsignor.

"All my life I've known him…as a friend and mentor. How do you know him?" the doctor asked Lovie.

"We believe that he kidnapped and enslaved thousands of people and used them in horrible laboratory experiments."

"Here we go again," said the monsignor. "The misdeeds of Dr. Scorn will not…let go…of me. Young lady, for your information, those kidnappings happened…before my science team was assembled."

"Don't even try that. I got the proof right here," said Lovie, throwing the log-book at him, but the monsignor caught it as though he was an athlete.

Lovie issued a challenge, "Go ahead. Read the truth if you got the stomach for it."

The monsignor obliged. He read the first few pages and stopped. "This is horrible. It confirms what I've heard, but these incidents happened before my watch."

His denial infuriated Lovie. "The Inquiry killed my brother—and his best friend—trying to get that book. Now you say you don't know anything about what's in it?" she shouted.

Raquel said, "The Cyclops's activities were entered chronologically, so you should start with the last entry and read forward from there. By the way, when did your science team start working for the Pluralists?" she asked.

"The middle of 2013," the monsignor replied, as he hurried to read the entries on the last few pages. A moment later, he stopped and closed the book in disgust. Then he sat on the edge of his patient's bed and stared at the floor in disbelief.

"It's nauseating, isn't it," said Lovie.

"I can't believe it...but the last three entries happened on my watch," he said.

"I'm not surprised," said Lovie.

"So far, my name has not been mentioned. But I knew the captain of the Cyclops, and this is his handwriting."

"Be careful, don't you know what you are admitting?" Father Paul asked.

"I believe that the events and the dates are accurate—which means the Inquiry did this without my knowledge," said the monsignor. He turned to face Lovie's hostility. "Ms. McGowan, if there are any additional reasons for you to distrust me, maybe this is the time to reveal them."

"Trust you? I can't afford to do that because you might be in league with my brother's killer," said Lovie.

"That's not a very nice thing to say to a holy man," said Raquel.

"I understand but I can't take the chance," said Lovie.

"Translated, that means you are guilty until proven innocence," said Father Paul.

"I see," said the Msgr. "Ms. McGowan, I will agree to be guilty if you will agree to let me know the moment the truth tells you otherwise."

The eyes of the heroes went from one to the other. They wondered if they should trust the holy man dressed in a bishop's choir cassock and a fuchsia sash. "You have my word," said Lovie.

"Good," said the monsignor, crossing his heart. "…there is one other matter. I want to assure you that the Pluralists are God-fearing people whose only mistake was trusting Dr. Scorn."

"I can vouch for them," said Father Paul, "but why did you get involved with their science project?" he asked.

"I'm a scientist and a man of the cloth. That may sound like a contradiction but it's not," said the monsignor.

"Did you ever consider what you would do if you discovered something that contradicted the holy Bible?" Father Paul asked.

"I don't worship the Bible I worship God Almighty. All my life I've straddled the worlds of theology and science. As a result, I have come to believe God is the truth and he can withstand the most rigorous scientific tests."

"But why did you join them," Father Paul asked.

"Numerous scientists have searched for the origin of humanity by getting on Evolution Road and traveling back in time. In contrast, the Pluralists wanted to travel in the opposite direction and so did I," said the monsignor.

"It's a good thing he did," said Dr. Kruel. "After the Gender Merger, the anatomy of the survivors became less specialized...but better prepared to keep human genes in the race."

"Is that why the genes merged?" Raquel asked.

"I wish I knew...we all do," said Dr. Minh. "Unfortunately, Mother Nature is the only one who knows why she started this merging process."

"What we know is this. The Gender Merger is in our future and we'll probably lose most of the world's population unless we inoculate them with the Serum," said the monsignor.

"So that's what you've been doing?" Father Paul asked.

"That's what I've been trying to do," said the monsignor.

"You've given all of us a real chance," said Raquel, "But I thought you were in Rome. What brought you to the Island?" she asked.

"I heard that you had hacked several computers using my password. So I thought you'd come here next," said the monsignor.

"That's good figuring," said Father Paul.

"I'm not the only one who reached that conclusion. The Newbies captured several members of the Inquiry trying to sneak onto the Island. I believe they came here to get the logbook. So you'd better hold on to this, young lady," he said, handing the logbook back to Lovie.

32. Face to Face

Saylor and two other Newbies entered the hospital ward shoving two prisoners ahead of them. "This is Dr. Scorn and Konnerman, his assassin. You may want to question them after you hear what they've been doing," he said.

Rage rushed through Lovie when she laid eyes on members of the Inquiry for the first time. She squeezed her fists and forged their images into her memory forever.

Scorn was dressed in a military camouflage uniform. He'd darkened his face with charcoal dust, but the stare he gave Lovie was darker. Konnerman was wearing a garment that displayed her muscular physique. One glance and Lovie remembered Muggs saying— a muscle-bound person had snapped Mike's neck. The two women locked eyes and stared at each other with looks that said, *'Till death do us part.'*

Suddenly, the sound of Saylor's voice brought Lovie's attention back to the more pressing issues. "The Inquiry is stirring up trouble in Washington, because they don't want any of us to leave the Island."

"You lost me there," said Father Paul.

"Saylor is referring to the 'emergency plan' we drew up in case control of the serum fell into the wrong hands," said Dr. Scorn, with a devilish grin.

Father Paul left little doubt about his attitude toward Scorn when he said, "I'll be glad to see your horrid days come to an end."

"Thine days are numbered with mine," uttered Scorn.

"What are you talking about?" Father Paul asked.

"...and for once in your life try telling the truth," Raquel suggested.

Scorn seemed eager to dish out the painful news before the Newbies revealed it. "The Inquiry called for an emergency session of the UN Security Council. As soon as they convene a quorum...they will decide whether or not to nuke this island."

"He's speaks the truth," said Dr. Anh. "They're lobbying hard for a quick strike. They want all evidence against them destroyed—which includes the log-book, the serum, and all of our people."

"But the serum is the hope of our children. Destroy it and they destroy man-kind," said the monsignor, who appeared drained of hope by the calamity.

"How can we protect these people? Surely there's a plan for this situation?" Father Paul asked.

"The Newbies can stop the bombing dead in its tracks any time they want to with ELF," said the monsignor.

"I thought ELF was a neural network for communicating with each other," said Lovie.

"...unless they turn up the intensity. Then it can be lethal," said the monsignor.

"Does it have to be lethal...?" Father Paul asked.

"How about an intensity level somewhere in between," said Lovie.

They could use ELF to disrupt Earth's communication systems, making it impossible to deliver the bomb," said the monsignor.

Raquel asked, "Is that how you knocked out my daddy's guards and the thugs on the ship?" But Saylor ignored her.

"Probably," said the monsignor. "...but they are pacifist. They won't even discuss ELF. They believe that talking about it draws the attention of people who would use it as a weapon."

"That sounds intelligent but we got an emergency on our hands," said Lovie.

There was a moment of silence that seemed endless until Dr. Anh shared an ELF message she had just received. "I have good news. The Security Council will not have a quorum until tomorrow."

"We got a break. What we do in the next few hours may well determine our destiny," said Father Paul, breathing a sigh of relief.

"We need to tell the world," said Lovie. "The people have to know about the Gender Merger and the serum so they can decide their own destiny."

"It's for sure we can't leave our fate in the hands of UN delegates...so what's next?" Raquel asked.

"Come with me. I've got an idea," said Saylor, waving for everyone to follow him in the direction of the studio.

33. The Webcast Studio

The Pluralists had completed the studio six years before with the help of the Newbies, and the results were awe-inspiring. It was a virtual platform designed for rapid production and delivery of religious services via the Internet, but no one seemed to be on the premises.

Saylor yelled, "Hello...hello... Warner? Is anyone here?"

"Who wants to know?" asked Warner, rolling out from beneath some equipment on a mechanic's dolly.

Saylor appeared glad to see him. "Folks, this is my good friend, Warner. He's the studio manager—the person who can give the people on this planet a voice in all this." Warner bowed to the visitors and Saylor went on, "We were born the same year, but you can't tell that from the way he's aged." He smiled and patted his friend on the back.

"He's telling jokes, so he must want a big favor," said Warner. "And my answer is yes...what do you need?"

"Do you remember the teleconference we discussed some time ago? Well, the time has come," said Saylor.

"You most certainly picked the right place for communicating with the world. I'll give you a quick tour, but watch your step. We've got a lot of cables and props scattered on the floor," said Warner, waving for the group to follow him.

"We are a content delivery network, so we're prepared to host big events like the one you have in mind. We have all the standard technology—the green-screens, social media integration—and our transmissions are compatible with every type of mobile device. We have all the standard capabilities but we've gone a step farther."

"...in what way?" Saylor asked.

"We tweaked our equipment with ELF technology. First, we used ELF to create a virtual studio unlike anything else on this planet."

"Meaning...what?"

"We mess with the senses big-time—and I'm talking about more than audio and video. What I'm saying is...we can deliver sensations in the form of smell, taste, and touch over the airways."

"I'm listening..." said Saylor.

"If a scene requires something unique, we can create it for you. If your audience is watching a person eating liver and onions smothered in gravy, served with rice and croissants, we will record the flavors and aromas on separate tracks and transmit them along with the usual audio and video tracks."

"Wow, how do you like those apples?" Father Paul asked, smacking his lips and rubbing his stomach.

"Our stagehand crew includes flavorists as well as technicians for lighting, sound, and special effects. Our network and websites are protected from viruses and swarming. There's satellite linkage and Chatter's language translation software is ready to go on your mark."

"Okay," said Lovie. "Now all we have to do is convince the world to visit this website in ah…what do you say…twenty-four hours?" Lovie asked, checking her watch for the time.

"Sounds about right," said Dr. Anh.

"During which time we have to produce one heck of a telecast," said Lovie, mentally measuring the size of the project as she spoke.

"Let them handle it," said the monsignor, nodding his head toward the Newbies.

Lovie liked the idea of getting some help. "…Saylor is this similar to what you and Michael had in mind?" she asked.

"Other than the atomic bomb threat…I'd say yeah."

"Then it's still your show…if you want it."

"This is a wonderful way to introduce my people to the world," he said.

"But do we have enough time?" She asked.

"We've had our backs against the wall several times before. Come in here. I want to show you something," said Warner. He opened the door to a large kitchen and flicked on the lights. "This kitchen is set up to prepare meals for crews having to work long hours to meet serious deadlines. Over here we have accommodation for the victory celebration afterwards. We usually celebrate our successes by rolling in chilled bottles of vintage wines on this cart. Look…there's the corkscrew ready to uncork the bottles."

"I figure we should introduce the Newbies first. Then we should warn them about the Gender Merger and our plan for managing it. Finally, we will tell them about the implications of the upcoming UN vote," said Warner.

"Lovie —we'll have something ready shortly," said Saylor. Then he and Warner began planning the first World Internet Conference

34. Wake up the World

The next morning, the Newbies launched an electronic banner programmed to appear on every video monitor across the globe at precisely 9:00 a.m. local time.

Mumbai, Beijing, and Tokyo were the first major cities to have their morning television programs interrupted by an announcer who read the banner as it scrolled up their viewers' TV screens. Network managers went berserk trying to stop it but they were helpless. The audience could not see the person reading the banner; however, there was something about his uncanny voice that compelled their attention:

Good morning, *Earthlings…*

> *Please don't be alarmed. There is nothing wrong with your TV screen or your computer monitors. Your program will reappear momentarily. We are Newbies—a new Human Life Form living on your planet. You are probably wondering if this is a hoax but we assure you it is not. Naturally, you want to know how we got here. Are we Friend or Foe? Are we Alien, Gay, or Straight? You are welcome to find out. You can join us in a live videoconference by simply turning on your TV or monitors tonight at 8:00 p.m., Greenwich Mean Time. Alternatively, those who miss the conference can learn all about us from a book available on our website https://www.newbiewebsite.co*

This television interruption followed the morning Sun as it moved across the earth. Wherever the clock struck 9:00 a.m. the Newbie Banner and the announcer took control of local TV sets and computer screens. The message was unavoidable since it displayed on every network.

Broadcast and cable stations rushed to be the first to discuss the phenomenon. They reran the banner and analyzed its meaning over and over, refueling the commotion.

Berlin TV reported, "Talk of this phenomenon is spreading like a wildfire around the world. Some people reportedly sat in front of their TV sets before nine a.m. and waited for the banner to arrive…then they recorded it."

Hong Kong TV reported, "This does not appear to be a hoax. World governments found the website, but they cannot find its internet service provider, so the site cannot be shut down."

Paris TV reported, "Members of the United Nations quietly raised their military alert status without alarming the citizenry."

Washington TV reported, "Word of the videoconference, fueled by human curiosity, went viral in record time. It spread exponentially all over the world. Families and friends gathered on their farms, at their workplaces, and in their homes to see it for themselves."

Eerie things begin to happen as the time of the webcast drew near. For one thing, the Internet came to life. The instant a computerized device connected with the Net the Newbie website took control of it and displayed a countdown clock.

The clock showed there were only eight...seven...six seconds to go before the World Internet Conference began.

35. The World Internet Conference

The World Conference started at 8 p.m. sharp. It began by putting the viewers in virtual outer space with a glimpse of Mother Earth off in the distance.

The audience saw Earth as a beautiful blue sphere with swirling white clouds moving across the heavens. Mother Earth got bigger and bigger as the eye of the camera brought the audience close enough to recognize the continents.

Earth's rotation seemed to stop when the lens looked down upon the Pacific Ocean and made Newbie Island visible. *Click-up* was the sound heard around the world as the camera closed in on the Island at a speed that thrilled the audience like a hi-speed roller coaster ride. To enhance the excitement, the Newbies used ELF to ensure the viewers felt the rough bumps of a real ride. The audience felt threatened, confident and excited at the same time.

When the eye of the camera got close enough to reveal the buildings on the Island, the camera homed in on the webcast studio. Then it took the audience up the steps, through the front door and continued until it reached the studio where Lovie waited to greet the world.

36. Lovie's Greetings

"Welcome to Newbie Island and to our Webcast Studio. I'm your moderator, Lovie McGowan and as you can see I'm an Earthling. I know you want to meet the Newbies as much as they want to meet you. And we're gonna' do precisely that but first we need to answer some general questions for everyone.

"So what better way to start than with the scientist who brought the Newbies here? Ladies and gentlemen, I give you priest—and scientist—Monsignor Mancini."

The studio audience applauded as the eyes of the world watched the monsignor step up to the podium. His backdrop was a huge floor-to-ceiling TV screen where a picture of himself, Dr. Kruel, and Dr. Minh was paused.

Mancini began, "Evolution has always been a controversial subject, because a large portion of the evidence is equivalent to hearsay from dead creatures. I probably look real, real old to the youngsters watching this webcast, so I want to assure them that I'm not old enough to speak with any personal knowledge about the dinosaurs and the ancient cavemen. Believe me, children, I was not there."

The monsignor waited long enough for the audience to enjoy his humor. Then he continued, "However, I *was* there to witness and record every marvelous step of the Newbies' evolution. The doctrine you are about to see is from my records. It is the truth, organized and presented to bring you up to date so that we can have a productive Q & A session."

"Follow me," said the monsignor as he walked across the studio toward the big TV screen. Once there, he looked up and saluted his colleagues. Then he pointed at his own image and said, "If you have any doubt about what you see on the video, remember—I'm the good-looking guy standing in the middle."

He pressed *play* on his remote control. The video began, and the world watched a documentary called *The Origin of Newbies*.

The world's viewers watched the fascinating story from the beginning to the end. As expected, a divided audience ensued. The mistreatment of gays made some viewers feel guilty because it reminded them of how they had looked the other way when they saw gays mistreated. To them, the need for personal reconciliation was obvious. So they made promises with their gods to make amends. However, the majority of the audience were straights who found gay behavior despicable. A wide gap existed between the two groups, and the straights did not want to close it.

The Gender Merger was the only issue that unified the audience, because it was a threat to everyone. The world audience considered it the number-one issue and they wanted action taken right away.

37. Q & A

When the last scene of the video concluded, Saylor walked up to the podium, buck naked. "My name is Saylor, and I'm one of several Newbies selected to answer your questions." After the viewers had a chance to check out his anatomy, he put on a robe and waved at several more Newbies to join him in front of the cameras.

Lovie said, "We have assembled a cross section of Newbie society for our panel. They are grandparents, parents, and teenagers. The monsignor, his science team, and Father Paul will join the Newbie panel to answer your questions.

"Does anyone on the panel have any last-minute comment before we begin?" Lovie asked.

One of the Newbie physicians stepped forward, "I'm Dr. Anh. I want to warn you about the loss of privacy. For those of you living in countries that do not allow you to talk freely on the air...you should remember that everyone in the room with the person who talks with us will be seen and heard as well."

Lovie said, "Since the subject of law enforcement has come up, I want you to see who the Newbies captured trying to sneak onto the Island." Then a picture of Dr. Scorn and Konnerman sitting in the studio, under arrest, appeared on monitors all over the world.

"Father Paul, I believe you have something to say before we start," said Dr. Anh.

But Father Paul couldn't remember the topic. He put his hand beneath his chin and thought aloud, "Darn it, I had something to say…" he mumbled and the people of the world laughed at his predicament.

Lovie said, "Languages, maybe…?"

"Right you are!" said Father Paul, who retrieved from his watch pocket a neatly folded note he had written to himself. "Our website uses Chatter language conversion software. So go ahead, express yourself—and the audience will understand you."

Lovie said, "If you're ready…we are ready. And our first surfers are from Ottawa, Canada. Are you there, Ottawa?"

Question 1: Are You Friends or Foes?

"We are here and my name is Ethan. My sister Heather, my friend Geoff and I have questions for the Newbies."

"We are ready," said Dr. Anh.

"I can't speak for the rest of the world…but you intimidate me. I feel threatened maybe it's because I don't know you, but that's how I feel."

"Thanks for your honesty, said Dr. Anh. "Do you have a question?"

"Sure. What do you say to other people who are as uncomfortable as I am?"

"If you watched the video about our origin you know that you are our ancestors and that harming you would also harm us. Wouldn't it?"

"It sure seems that way," said Ethan, feeling somewhat relieved.

Heather said, "I'm worried about the Gender Merger. Every ounce of our hope is relying on the serum. How's it coming along?"

"The serum has passed every test we can imagine," said Dr. Minh.

"Tell us about the most recent test. How did it go?" Heather asked.

"It was exciting," said Dr. Minh and her smile got brighter the more she expressed her delight. "We made another trip to the tunnel and this time all of our volunteers made it through the entrance without incident."

"...which means what?" Heather asked. "I don't get the point."

"None of our volunteers carried the gay DNA strand *until* we injected them with the serum," said Dr. Minh.

When the words struck home, Heather's happiness exploded with a big *Hooray* as she jumped up and down in jubilation. "Then the serum works?" she yelled.

"We don't have a guarantee that they'll exit the other end of the tunnel," said Dr. Kruel. "But isn't it wonderful to know that hope is alive and kicking?" he exclaimed.

Geoff, the third person from Ottawa, asked, "How do Newbies handle conflicts with each other?"

"We don't have a great deal of experience with it," said Dr. Anh.

Troi, one of the Newbie teenagers said, "We are competitive. But we concentrate on achieving objectives rather than preventing others from achieving theirs."

"Effort previously spent on competitive squabbles is redirected toward mastering knowledge that ensures our survival," said Saylor.

"Like what?" Geoff asked.

Abe, the other Newbie teenager, answered the question. "We've learned how to erect acoustic fences to steer rainy weather to drought stricken areas or we can send rain to drown out forest fires."

"...our engineers have learned how to avoid friction and how to store inertia for later use," said Dr. Anh.

"Your achievements raise my confidence in the serum, "said Ethan.

"I hope we calmed some of your fears," said Lovie. "We have to go now. We have new surfers from a medical university near Innsbruck, Austria. Are you there, Innsbruck?"

Question 2: A Question of Ethics

"Yes, we're here, and thanks for the opportunity to join this historical event. I'm Dr. Hans Gruber, along with my colleagues' doctors, Steiner and Weber. Our questions are for the monsignor."

"I'm ready," said the monsignor.

"Sir, you've opened a vast frontier for medical science to study and you've created an equal amount of controversy. So I have two quick questions: First, did you violate your Hippocratic Oath? And second, do you believe you committed crimes against humanity?"

"My answer is no to both questions. You will learn this when the facts unfold before the World Court."

Dr. Weber asked the next question. "Did your science team adhere to the evolutionary path Mother Nature was taking?"

"We believe so. We did our best to follow the genes."

"Ah, but you don't know for sure...do you?"

"We are not one hundred percent certain, if that's what you mean," said the monsignor.

"Well, this most certainly creates a problem," said Dr. Steiner. "Anthropologists start with a few bones and track our ancestral history backwards as far as there is a trail. If they lose the trail there is no damage. But in your case—"

Dr. Weber interrupted Dr. Steiner and said, "…which brings us to my question. You, sir, were pioneers—which mean there was no trail. So what did you do when you came to a juncture requiring a decision? Did you always go to the left, the right or did you play God?"

The insult cut the monsignor deeply, causing him to grimace, but he remained calm. "We learned that human intelligence starts a lot earlier than we imagined."

"What do you mean?" Dr. Weber asked.

"We learned that rational thinking occurs at the genetic level. Your genes perform all kinds of amazing feats. They think, communicate, and influence natural selection. So in most cases, our science team simply followed the genes."

"…in most cases?" Dr. Gruber asked, hoping he'd found a crack in the monsignor's explanation.

"An exception would be our reliance upon probability theories when we sped up evolution. However, the people nurtured in the process are here, and thus far, every indication is…our assessment was accurate."

"Yes, but you did this at the risk of altering the human race immeasurably. Surely you agree that some things are not ours for tampering? " Dr. Steiner asked angrily.

"God will be my judge and time will tell if my actions made our circumstances better or worse," said the monsignor with a firm tone in his voice.

"Thank you, doctors," said Lovie, bringing their talk to a quick close before it overheated.

She looked at the studio monitor that revealed the next surfers and saw three people sitting on a veranda, watching the webcast on their PenSet's monitor.

"It looks like we have a family from India. Are you there, Mumbai?" She asked.

Question 3: The Gay DNA Strand

"This is Mumbai," said a well-dressed man wearing an embroidered outfit. He and two ladies were watching the webcast together.

"I'm Nihal. This is my wife, Celina, and our daughter Shanta. We have a question about the gay strand."

"Please continue," said the monsignor signaling for Doctors Minh and Kruel to come and stand next to him.

"We don't understand what the gay strand does for people. Can you clarify this?" Nihal asked.

The monsignor said, "You should start by understanding that the 'tunnel' *is* the time and place on 'Evolution Road' where the unique features of our male and female genes attempt to merge."

"Gotcha," said Nihal.

"Also, keep in mind that it took more than a thousand 'laboratory years' for the genes of our volunteers to reach the end of the tunnel," said Dr. Minh.

The monsignor said, "We recorded everything. It was marvelous to witness the role the gay strand played for those lucky enough to be carriers. It was fascinating."

Dr. Minh continued the explanation from there. "The strand understood the complex needs of the male *and* female genes. So it made sure that the carriers had some portion of their genetic changes in progress before they entered the tunnel."

"Getting a head start was extremely important because there was a shortage of time," said Dr. Kruel. "The Strand sequenced the genes and selected the characteristics to be passed on to the offsprings."

"Then it guided the offsprings out of that frenzied tunnel to safety," said the monsignor.

"It was amazing but unfortunately the 'straights' were not so lucky," said Dr. Minh.

"This was a terrible time for straights, because their genes were not prepared," said Dr. Kruel.

"It was chaotic," said the monsignor. "The straights' genes had done nothing in advance. Consequently, the demand for them to start merging happened too fast for them to adapt."

"It was horrible," said Dr. Minh. "At first we heard these disturbing wailing sounds. They were faint at first but it didn't take long to recognize we were hearing the cries of the 'straights' expressing their pain and suffering."

"...what happened to them?" the daughter asked, hoping for a happy, fairytale conclusion.

"They no longer exist," said the monsignor.

A sense of desperation was clearly on the face of Nihal's family, "If we are inoculated with the serum, will we become carriers?"

"The serum will give your genes and that of your offsprings passports through the tunnel, "said Dr. Kruel.

"When can we get the serum?" asked the wife, who by now was well convinced.

"We'll work out the details with the World Health Organization and they will let you know," said Dr. Anh.

I hope we've answered your questions Mumbai because we have surfers calling us from Great Britain," said Lovie. "Can you hear me over there?" She asked.

Question 4: The Serum Debate

About forty people were watching the webcast in a London Country Club. They were debating issues about gays, Newbies, the Gender Merger, and especially the risk associated with taking the serum.

Nigel Sheffield, a frequent patron of the club said, "I don't mean to be a difficult snob but straight people like me find homosexual behavior appalling. Frankly, I shudder at the thought of what goes on behind closed doors. Know what I mean?"

It wasn't as if Nigel needed any encouragement, but when everyone in the club cheered he continued talking. "You may call it hate, fear, disgust —or any bloody thing you choose. But God gave me my preferences the same as he gave gays theirs."

Ian was a bearded guy wearing large-framed eyeglasses. He sat in a far corner of the clubhouse, rocking back and forth on the hind-legs of his chair. He'd listened intently before he spoke. "I want to talk about that serum of yours. As I understand it serums are made from blood —is that correct?" he asked.

"I agree with that," said Dr. Anh.

"Okay, so tell me. Where do you get the blood used to make the serum you intend to inject into these people?" he asked.

"Blood extracted from Newbie placentas is used to make the serum," said Dr. Anh.

"You can't be for real?" said Liz. "Have you people lost your bloody marbles? I don't fear evolving over an extended period of time and slowly becoming a Newbie. But if hurrying to get ready for the Gender Merger means I have to become gay first then forget about it—I want no part of this."

"I have to agree with my wife on this one," said Lloyd, pointing to Liz. "If I have to be injected with the serum to become a Newbie and it turns me into a puff I rather die the way I am Nothing personal of course but I just couldn't do it."

"The serum will not make you gay," said Dr. Anh.

"Will we act gay?" Nigel asked.

"Right now you are acting as though you don't trust us," said Dr. Anh.

"We're being precautious," said Nigel, who then turned to speak to all of his compatriots. "Everyone has to remember that these Newbies are not aliens from a distant world—they're test-tube babies brought to us by a bunch of nutty scientists."

"I'll kill myself before I let them turn me into anything remotely like a queen," said a guy with a burly chest.

"And that goes double for me," purred the voluptuous blonde draped all over him.

"Whoa, we need to slow down a bit," said Lovie. "Let's start by laying a few rules on the table. Number one, the serum will not make you gay. And number two, this is a volunteer program."

"So, if I understand you right there are two ways I can handle this situation," said Ian. "If I do nothing you say most of my offsprings will die because they can't make it through the tunnel without the Gay Strand. Or I can let you inject me with a serum that might turn me into a raging puff. I need another option!"

The barmaid spoke up. "For goodness sake listen to the scientists. I have two children at home who desperately need answers. We could have lived with many things that Mother Nature sent our way, but we managed to come up with all kinds of solutions. This situation is no different. If those Newbies have figured out a safe way for my children to make this transition, I'm gonna take it."

Maggie, a patron sitting at the bar said, "I react to homosexual behavior like I do when someone scratches a chalkboard with their fingernails. It makes my flesh crawl and by George it irritates the blooming hell out of me. That's why I can't stand for them to be around me."

A man dressed like a preacher hollered, "I say no to the serum! What'll happen if we interfere with what the Lord has in mind for us? What do you think we ought to do Father Paul? Are you listening to all this?"

"I'm here," said Father Paul. "Don't think you are an exception because you have conflict in your clubhouse. Tonight, families all over the world are sitting in their homes watching this webcast and I guarantee there is more than enough conflict to go around —that's why taking the serum is a volunteer program."

"Speaking of conflicting opinions," said Lovie, "—earlier today some students from an American university offered to let us eavesdrop on what they believed would be a heated discussion in their fraternity house. Maybe this is a good time to be a fly on the wall and listen to them. We will go there now—unannounced of course."

Question 5: Frat House Fears

College students had gathered in a fraternity house to socialize and watch the webcast on a big-screen TV. Beer, booze, chips, and dips were in abundance—and so were differences of opinion.

An extra-large fellow known as a campus bully tried to convince those sitting around him that Newbies were a race of schizophrenics, ricocheting back and forth between male and female identities, "I hereby charge all Newbies with being both a man and a woman!" he shouted as he raised his mug of beer up toward those looking down from the second floor.

An athletic female student dressed in a joggers' outfit and a towel wrapped around her neck took an opposing stance. "That's ridiculous," she said. "Newbies straddle the space between the two sexes; they're neither male nor female. They are Newbies."

Another student who was using a calculator stopped to say, "I agree. It's simple. So let's not complicate it."

The bully rejected both of their explanations and countered with one of his own, "Newbies are no different than gays. They're like a contagion spreading every chance it gets. I tell you...more and more of 'em are born every year. Right now they probably represent twenty-five percent of the population."

"Hush, you are frightening some of the guests," his girlfriend pleaded. But he ignored her and spoke to the woman on the second floor. "…Ma'am, if you think this is frightening, what's gonna happen when the gay percentage increase to thirty or forty percent…huh? You ever think about that? Tell me, what percent is your boiling point?" He asked.

"We have enough of 'em already," someone said in support of the bully which encouraged him to say,

"All I'm asking is a simple question. Is it better to handle the gay takeover now or should we wait until we're outnumbered?"

"Oh yeah, that makes a lot of sense," was a comment from the second floor. "I can see us taking up arms and starting wars within our own families. That certainly makes a lot of sense."

"The gay agenda is a big threat to religious teachings," said the Bully.

"What about God? Do you think he has a hand in your so-called 'gay agenda'?" the athletic woman asked the bully.

"Being gay is a personal decision. God has nothing to do with their being homosexuals," he replied.

"Oh yeah, and when did you decide to be straight?" The athletic woman asked.

"I…I didn't, nobody asked me," said the bully.

"Precisely…and no one asked the gays either. For your information, Almighty God decided for all of us," said the woman.

"My, gosh," said the student with the calculator. "Our understanding of gay issues is just as fuzzy as our understanding of economics."

Lovie stepped in front of the camera and spoke to the world audience, "I have to agree with the student using the calculator. Anyway it's time to greet our next surfers who comes all the way from Italy. He's a dad, and his son is a patient in a hospital near Rome.

"Mr. Covelli, we are here. Do you still want visitors?" Lovie asked.

Question 6: Gay Bashing

Joseph Covelli wore the haggard look that comes from sleep deprivation. He looked older than his forty years and his voice sounded very tired, "I invited you to visit me and my son Vitto so you could see what a cowardly bunch of bullies did to my boy," he said.

Unable to hold back the tears, Mr. Covelli extended his arms and welcomed the camera to scan the hospital room so the rest of the world could bear witness to the assault upon his son.

Monitors of various types displayed the vital signs of a seriously injured teenager fighting for his life. Vitto was unconscious. The tiny lights on the monitors were flashing while an IV drip, hanging above the boy's bed, fed fluids to his frail body. Vitto's face was bruised and swollen and he had a deep gash in his lower lip. One arm was in a cast and he required the help of a breathing machine to stay alive.

"What happened?" Lovie asked, already tearing up.

"The other students don't like him. They say he's gay, and I say he's my son."

"Did this happen today?" Lovie asked, remembering similar incidents with Mike.

"Today, yesterday—all the time, they bully my boy."

"Do the bullies attend his school?" Dr. Anh asked.

"Bad boys and bad girls attend his school."

"Why did they do this?" Abe asked.

"The bullies harass kids about everything. They make fun of the child who stutters. They find fault with the children whose appearance is different—be it hair, clothes, color or weight…which we all know is a pile of crap. If a child is different, they give him a beating instead of help. Now place being gay on top of that and you get an idea about Vitto's life."

"Would changing schools help?" Lovie asked.

"…the problem is everywhere. Ask the families in your audience. I bet they know what I talk about."

"Vitto cried this morning. He said he didn't want to go to school, but I forced him to go because he'd done so much homework on his school project. Now look at what's happened."

Lovie said, "I will get you some help. And that's a promise."

"It's too late for help. Vitto does not want to wake up. He has no desire to live such a horrible life…would any of you, if you were him?"

No one attempted to answer, so Mr. Covelli looked into the camera and asked, "Is there anyone on this panel willing to change places with my boy?" He asked again but still no one answered.

"Ah…now you see why he'd rather kill himself than continue living." said Mr. Covelli.

"Death is not a desirable outcome," said Dr. Anh.

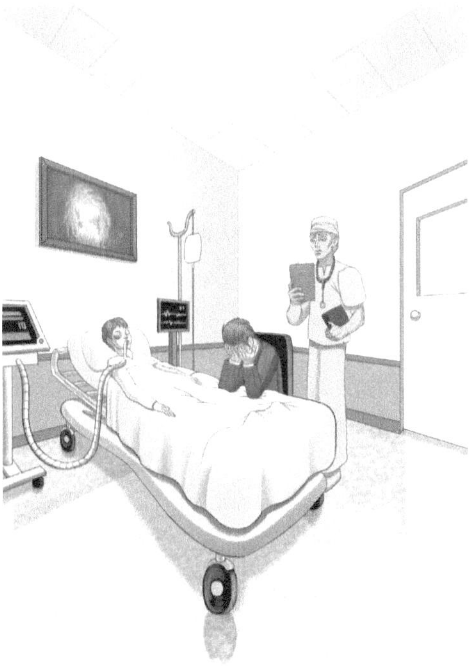

"You think I don't know that? Maybe I should put an end to this. I can kill my son and that would end it for him. But I think I'm gonna kill a few of those bad kids first. I'll do this for my Vitto. He is such a good boy."

Mr. Covelli spirit heightened when a physician entered the room. "This is Vitto's doctor," he said, pointing to the man checking the boy's charts.

"How is Vitto doing?" Dr. Anh asked the doctor.

"We've done all we can, but it will be several hours before we know anything for certain," he said. "I've watched portions of your webcast between patient visits. Tell me, how do Newbies handle bullying?"

Dr. Anh said, "Our teenage panelists have a couple of suggestions. Troi will you speak first?" she asked.

"If I was a person looking for a mate I would steer clear of anyone known to be a bully," said Troi.

"...and your reason...?" Dr. Anh asked

"A bully might bring his or her cowardly behavior into our relationship," said Troi.

Abe, the second Newbie teenager said, "Bullies are cowards trying to do bad things in secret. Exposure is their worst enemy. I suggest that the school take a page from the teachings of Father Paul and implement a variation to his '3-P' program—record bullying incidents on cell phones or school cameras and share it with the children's *Parents*, the *Police* chief and the school's *Principal*.

"Parents can help by asking the police chief to setup an email address to receive the recordings," said Troi.

"Those sound like great ideas," said Lovie, staring into the studio camera. "So I'm asking anyone out there in the audience to tell me what you think of the suggestions."

After a moment, Lovie said, "We've got a reply. From where are you calling? " she asked.

Question 7: Teen Suicides

"We're in Salt Lake City, Utah. There are six families visiting me," said a woman sitting on a couch.

The man holding her hand said, "All of us are parents of children who committed suicide. We decided to watch your webcast together because we're determined to do whatever is necessary to end the suicides and the bullying that cause them."

A mother whose eyes appeared red from crying said, "We are Mormons, and oh yes, we are the same Mormons who years ago, left the greats states of Missouri and Illinois to escape the kind of persecution some of us inflict upon gays, today."

The first mother said, "Children must learn to stand together and report bullying before it pushes all of them over the edge."

One of the brave mothers shared her daughter's tragic story, "I'm ashamed to admit it but my daughter Megan was a member of a girl's gang that bullied other kids. That is until her best friend's boyfriend took a liking to her. Megan was innocent but the gang accused her of stealing her friend's boyfriend. The gang turned against her. They rolled up magazines and beat her on the head. They called her a dyke and threw their lunches at her. Megan discovered that the gang used lipstick and drew obscene pictures of her on the bathroom mirrors. This upset her something awful so she decided to erase the pictures.

The gang caught Megan wiping the mirrors clean so they beat her up and banged her face up against a mirror over…and over so many times it cut up her face something terrible. Megan could not handle the scars and stitches all over her face so she escaped their harassment by killing herself."

A brave couple who seem to give each other strength spoke about their son's ordeal, "When our son Billy was eight he was terrified of the dark. We could not leave him in his room alone because he thought the boogie-man was hiding amongst his toys in the closet," said the dad.

"It was terrible and we were desperate for a solution," said the mom. "So we bought him a toy; a fierce-looking teddy-bear to give him courage. No one could have been more surprised and pleased when it actually worked. He called his toy Boogie-Bear and they became inseparable."

"He slept with it all the time," said the dad. "He treated Boogie-Bear as if it was his personal bodyguard."

"We suspected Billy was gay but we didn't know for sure," said the mom. "We also had no idea that Billy angered the school bullies when he took Boogie-Bear to class."

"The last straw occurred when the bullies beat him up, ripped off Boogie-Bear's head and ran away with it. Gee'meneez, it was just a toy so why did they have to take it from him?" the father asked.

"A few days later Billy buried Boogie-Bear's headless body in our backyard and hung himself from a tree next to the grave," said the mom holding onto her husband for support.

"Joyce and I read Billy's suicide note and I swear we had no idea how difficult life was for him. We are angry with the school for not telling us about the bullying incidents. So we can identify with Mr. Covelli's plight. We also think a program like '3-P' will expose the bullies and place accountability squarely on the shoulders of those responsible for correcting the problem."

"Wow," said Lovie, "Remind me again…who gets notified under the '3-P' program?" she asked.

"The Parents, the Police and the Principal," the Mormon's parents yelled in unison like a group of cheerleaders.

"No one could have said it better," said Lovie. "We love you but we have to go. Our next surfer is a mom who asked us not to reveal her whereabouts. Are you there ma'am?" Lovie asked.

Question 8: An Arabian Mother

A woman dressed in black attire that covered most of her body turned to face the camera embedded in her monitor. She wore a veil that shielded everything but her tearing eyes. Her voice was soft and it trembled when she spoke.

"My question is for anyone who can help me. My daughter is a lesbian and I worry about her safety. My government claims that we don't have any gays in our country but last month they dragged several men from the house across the street and accused them of being gay."

She pointed to the window, "I stood there and watched them. I saw the police whip those helpless men unmercifully out there, on the streets, for everyone to witness their humiliation."

Speaking out against her government was a difficult and brave act. It was enough to force her to pause and gather her composure, "The police held the men down and injected them with an anti-gay hormone ordered by our government. Then the police threw the men into a truck and drove away."

"Drove them where?" Lovie asked.

"Who knows? They are the police which mean I have nowhere to call for help. I don't know why I took pictures because I have nowhere to send them," she said in the midst of her sobbing and confusion.

Dr. Anh said, "I will check to see if there's a way to get you and your daughter out of the country. We know how to reach you and we will get back to you."

"At least that's a start," said Lovie whose eyes were as watery as the unnamed mother's eyes. Then Lovie shook off a cold chill that tried to invade her body. "Our next surfers are from France. Are you there, Marseilles?

Question 9: The Human Race

"This is Marseilles; we are here. I'm Dr. Claudia Farina, and this is my associate, Dr. Léon Rameau. We have a question for the science team,"

"What is your question, Doctor?" the monsignor asked, stepping forward like a medieval knight about to accept a challenge.

"We would like to revisit Dr. Steiner's question when he suggested that you may have altered the human race immeasurably. Do you have an answer for that?"

"Frankly, I'm not sure if I know what Dr. Steiner meant when he used the term 'human race,'" said the monsignor.

"Oh, come on, Doctor, please don't be absurd—" began Dr. Farina.

The monsignor interrupted. "Would you be kind enough to indulge an old man by answering a very simple question?" he asked.

"...I would be honored, Doctor," she replied.

The monsignor expressed his gratitude with a nod before he spoke, "Have you ever asked yourself what the term 'the human race' means? Think about it. What does it really mean? Does it *identify* who we are...or does it *describe* what we are doing? If I may, I suggest the world audience consider the question as well. Take a moment and ask yourself the same question. What does the term the '*Human Race*' mean?"

...And the webcast viewers did precisely that. The question ricocheted around the globe and the world became silent while the people thought about how they would answer.

Dr. Farina was the first to speak. "There is nothing simple about your question, Doctor.

Dr. Rameau, who had the voice of a bass singer, said, "Personally, I've always considered *'race'* to be a noun, because that's what I was taught. But now you suggest that it might be *describing what we are doing'* and that's profound."

"Have either of you ever noticed the size of the Oxford English Dictionary?" the monsignor asked, expanding his arms out to his sides to suggest an enormous book.

Dr. Farina watched him and said, "...I don't know, personally, but I get the impression that it's huge."

The monsignor agreed. "There are six hundred thousand words inside. So when some scholars place our ethnic groups into a category called 'races,' it makes me wonder about their motive."

"If the word *'race' is* describing what humans are doing, that would mean 'humans' are in some kind of contest...and that we're unaware of it," said Dr. Farina.

"Ooh la, la...I wonder if someone is trying to tell us we're in a race?" asked Dr. Rameau.

"I can't imagine that...a race for what?" asked Dr. Farina, her palms turned up toward the ceiling in dismay.

"Survival...what else is there?" the monsignor asked.

The French doctors' blinked as if they received a wakeup call and their intense stares showed they were giving the question considerable thought.

"...the idea that something might be chasing me is frightening," said Dr. Rameau.

"It's also confusing, since we humans are supposed to be the dominant species. We are...aren't we?" asked Dr. Farina, seeking reassurance.

"Human beings are not competing with each other in this particular race... and we're not being chased by another species. We are running from extinction," said a voice from the panel of Newbies. It was Dr. Anh, walking toward the science team.

"Extinction is an integral part of the natural order and it plays a significant role in evolution," said the monsignor.

"Which means that we're in one heck of a race and...it's for all the marbles," said Dr. Kruel.

"Running from extinction takes every new generation of humans into a new environment that inevitably requires us to use a brand-new set of tools in order to survive," said Dr. Anh.

"Can you give us an example?" Dr. Rameau asked.

"Mother Nature and acts of God can change the environment, and we have to adapt. During the Ice Ages, for example, some humans adapted by using fire as a tool to heat their caves and cook their meals," said Dr. Anh.

"Those who failed to learn how to make a fire became extinct," said Dr. Kruel.

"Here is another example," said Dr. Minh. "Modern man can change the environment by introducing new technology. If a tribe of cave dwellers learned how to make stone axes, the neighboring tribes had to learn how to make them too, or become extinct."

Dr. Kruel said, "…and you can apply that concept to other changes, such as using the wheel, harnessing electricity—or more recently, becoming computer literate."

"Notice how the tools get more complicated as our knowledge of science increases," said Dr. Anh.

"That's why education is essential if children are going to survive," said Dr. Minh.

Dr. Kruel, said, "I remember how several years ago, the Arab Spring protestors used the Internet as a tool to demand better governments. Their rulers were forced to adjust to the new environment or become extinct."

Dr. Anh said, "We are on Evolution Road, racing toward whatever—" Suddenly for some reason Dr. Anh stopped speaking. Then all of the Newbies turned toward Saylor as if they were waiting for him to speak.

He said, "I need everyone's attention. A moment ago, we received an ELF message stating that the United Nations has deployed an atomic submarine with orders to destroy Newbie Island."

Groans and moans filled the studio and despair was everywhere. The panel members began talking among themselves and the world audience overheard Lovie say, "The Inquiry is behind this. They want the serum at all costs, and if they can't have, it no one will."

"Don't they know that an atomic bomb will destroy the serum and humankind along with it?" asked Dr. Anh.

The calamity ignited fear and a wave of anger spread across the globe like a forest fire out-of-control.

38. The Protestors

The World's reaction was swift and clear. Broadcast and cable news stations rushed to the airwaves within minutes. They passionately welcomed the Newbies to the planet and pleaded for their governments to halt the bombing.

Some countries reported people expressing outrage in street demonstrations and several cities reported rioting. The conniving of the Inquiry rekindled the world's peace movement. Flotillas of protestors took to the sea with the intent to barricade Newbie Island and perish with it, if necessary.

Small groups of demonstrators merged with larger groups. Together they became hundreds, then thousands. They stormed the United Nations headquarters. The news media was there when those guarding the UN buildings told the crowd that the delegates who voted for the bombing had departed immediately after the vote.

39. The White House

The First Family and members of Congress were watching the webcast when they first heard of the UN's action. Immediately, the telephones at the White House and Congress began ringing constantly, and the United States military went on full alert.

"Mr. President—I regret to tell you, Sir, but the announcement made by the Newbies is true," said the White House Chief of Staff, "A United Nations submarine equipped with sea-to-air missiles is about to launch a nuclear strike upon the Island."

"Who could have authorized this?" the president asked angrily.

"Sir, the UN's Nuclear Test Commission has had this authority ever since nuclear testing was banished, but they never found cause to use it."

"...until today," chimed the president, as he struggled to get his arms around a growing challenge.

"What do you have for me?" he asked the Chairman of the Joint Chiefs of Staff (CJCS).

"Sir, the submarine is part of the military strike force placed under UN command by the same resolution."

"Jeez . . . how much time do we have?" the president asked.

"...two hours at the most," said the White House Chief of Staff.

"By orders, the captain will fire the missiles as soon as he reaches the launch coordinates," said the CJCS. "In the meantime, we have an armed satellite hovering over the Island."

"Can we contact the submarine?" asked the president.

"We're trying, but the captain has no idea the world is looking for him. Plus, his orders require him to maintain radio silence until after the mission is completed," said the CJCS.

The president appeared worried as this calamity closed in on the world. "The Inquiry seemed to have covered all the bases. Have the Navy locate the sub and do whatever is necessary to stop it from launching that missile and I mean everything," he ordered.

The president looked at his son, Spencer, sitting on the far side of the room. They had been enjoying the webcast together, but here was another wash-out of a father and son event. Now Spencer was over there by himself, still trying to submit a question to the Newbie panel.

"Yo...Spencer, if you get through, let me know. I want to speak with them also," said the president. Then he rejoined his staff.

Spencer resubmitted his question, but this time, he used the president's PenSet. After all, his dad had wanted to speak with the webcast panel as well.

40. What the...?

The appearance of the United States presidential seal on the studio monitors shocked everyone in the webcast studio. In fact, the sight of it gave them a serious case of paralysis until Lovie broke the spell.

"What the...what is that?" She asked. "Oh, my goodness, I think it's the White House. Are you there, Mr. President?"

Spencer was elated to see himself on TV, "Dad...hey, Dad! They're asking for you over here," he shouted, jumping and waving his arms.

The president turned to face Spencer and was astounded to see himself and his staff on TV, with the whole world watching. Never known to be camera-shy, the President quickly gathered his composure for the biggest audience of his life.

"Good evening, everyone," he said with a gracious smile. "I want you to know that the United States military is on full alert. We are attempting to contact the submarine, but apparently, the captain has orders to maintain radio silence."

"How much time do we have, Mr. President?" Lovie asked.

"...the captain will be close enough to launch his missiles in an hour and a half," the president said, confirming the calculations he and his staff had made.

"So...what's being done?" Lovie asked.

"The United States has established an armed perimeter around Newbie Island. It's for your protection and we will launch countermeasures at any threat that comes your way."

"What about the sub?" Saylor asked.

"If we don't hear from the captain within an hour, we will send instructions to the sub's computers to self-destruct. So, those of you out there in the audience, you can relax...we have the situation under control," he assured the world.

The wonderful news brought jubilee to the world. There was euphoria, and sighs of relief as the audience released its pent-up tension.

"Who is that standing next to you, Mr. President?" Lovie asked.

"This is my son, Spencer. He's seventeen, and he's been trying to submit a question to your panel of experts. I guess he succeeded."

"It seems as though we have at least ninety minutes—so in the meantime, how may we help you, Spencer?" Dr. Anh asked. But for some reason, the youngster hesitated.

Spencer looked to his dad for encouragement and got it. After inhaling some courage, he said, "I'm gay...and no one knows why. Did something go wrong with me?"

Spencer's admission and question numbed everyone and caused them to pause except for Troi, who spoke faster than his elders would have advised, "You're gay because you're a 'Transition Baby'."

"...I'm a what?" a baffled Spencer asked.

"You're a transition baby. It happens all the time," said Troi.

In all fairness, Troi believed that the 'transition baby' principle was a well-known scientific fact beyond Newbie Island, but was incorrect. The truth was...the world had never heard of it. So Dr. Anh rushed forth with some well-needed clarity.

"Have you watched the webcast from the beginning?" Dr. Anh asked.

Spencer nodded yes.

"Then what can you tell us about evolution?"

"Well...uh, I've heard that elephants come from the mammoths and that people come from apes, and that kind of stuff."

Dr. Kruel pounced on Spencer's answer, clowning like a court jester. "Of course, you realize that apes didn't go to bed one night and wake up as humans the next morning..." he asked, with antics that brought smiles to the faces of the president and his son.

"My teacher said it took a long time," said Spencer, laughing.

"Indeed it did. And during that long time, several versions of 'mankind' appeared and disappeared during the genetic merger that led to modern humans," said Dr. Minh.

Dr. Kruel said, "Each link in human evolution is separated by a transition period—the time when our genes evolve from what we are to what we are going to be."

Dr. Minh explained, "We call children born during the last quadrant of a transition period 'transition babies,' because they may be born with genes and organs that are well on the way toward completing the change…but sometimes, they may not be quite there."

"I think I understand," said Spencer.

"Okay, Mr. Wise Guy, here is a tough question," said Dr. Kruel. "If members of a species that's evolving from a very light color to a dark color had children during the transition…what color would the transition babies be?"

"That's a trick question," said Dr. Anh. "Remember, some transition periods last for thousands of years."

"I'd say…that the children would be any color between the starting color and where they are going," said Spencer, smiling and pleased with his cagy reply.

The monsignor and Dr. Anh approved his answer and the president gave Spencer a hug, "Spoken like a true politician," said the proud father.

Dr. Minh asked, "If that's the case, Spencer, what kind of changes would you expect to see from babies born during the ape-to-human transition?"

Spencer took a few moments to consider the question, "Ah…I think…the transition babies would walk upright more often—and they would have a lot less hair. Wouldn't you agree?"

"…with every succeeding generation," said Dr. Anh. And again, the panel members nodded approval.

"Okay, I have one more question," said Dr. Minh. "Modern humans' transition to Newbies involves significant changes to our sex organs, our glands, and the hormones they produce. What changes would you expect to see in the babies born during this transition?"

The thoughts in Spencer's mind spun like tumblers within a combination lock. Then finally he said, "Whew...there are enough changes going on to confuse the mind and the body."

"Excellent," said the monsignor. "...and even though Mother Nature tries very hard to synchronize those changes, inevitably they get out of sync."

"Are you saying that what people see on the outside of me might be out of step with my internal organs and hormones?" Spencer asked.

"...you answered your own question," said the monsignor. "...the reason you may feel or act gay is because your entire reproductive system is evolving. Under those circumstances some of your organs will certainly get out of sync."

Spencer's face lit up with surprise. "But I thought evolution was over...," he said, more as a question for anyone willing to answer.

"Darwin told the world that we are evolving and he was correct," said the monsignor looking into the camera. Then he gave the answer millions of families struggling with gay issues longed to hear. "If evolution was over we wouldn't have gays. Gay people are a transition step between Modern Man and the Newbies. Evolution is not over and the gay members of our society are the living proof."

The monsignors' profound statement brought tears of joy and redemption to the world audience. His words placed everyone into a state of thought akin to a trance until Dr. Kruel spoke.

"The Newbies are asexual," he said to Spencer. "Therefore, infatuation for members of the same sex will vanish as Modern Man evolves closer to becoming one of them. Just remember that Mother Nature has her own schedule so we must learn how to manage ourselves during this transition."

Raquel whispered to Lovie, "For the first time, citizens of the world have heard experts explain why some people are gay," Then she silently prayed that her mom was watching the webcast.

Although Father Paul appeared to be speechless he was far from it. "The question that asked why some people are Gay has nagged our society for centuries. It was the question families with gay challenges had asked their physicians and religious leaders repeatedly and received unconvincing answers...until today."

"Oh...yes and I heard the answer," said Raquel. "His explanation meets our criteria so it sits well with me. How can anyone ignore an answer that satisfies all of the known GLBT traits?"

"They can't ignore it," said Lovie, enthralled by a myriad of thoughts rushing through her mind—*doggone it! It was evolution all the while. The answer to Mike's question was staring right back at me the whole time, and I failed to see it. She wanted to holler to Mike and Seth to make sure they were listening, but somehow she knew they were all ears.*

The President said, "If I understand this correctly, what Gays thought was a curse was actually Mother Nature preparing them to survive the Gender Merger up ahead."

"You got it," said the monsignor.

Father Paul looked upward as though he were acknowledging a truth he had believed all his life. "Surely, the Lord works in mysterious ways...for now, those who society had looked down upon are the very ones chosen to lead us through that tunnel."

Spencer's elation confirmed his understanding as did the millions of families celebrating the good news. They didn't know if the monsignor was correct, but his explanation made more sense than any answers they'd heard previously.

"Mr. President, are you and Spencer all right with the monsignor's answer?" Lovie asked.

"This makes darn good sense to me," said the president. "I don't feel like a guilty parent anymore. "Thank goodness he's a scientist *and* a man of God."

Spencer said, "I'm glad I came out of the closet. Now that everyone knows why I'm gay, I feel like a huge load of stigma has been lifted from my shoulders."

41. The Grudge

Frustrated by the way fate twisted and turned against him, Dr. Scorn started fiddling with his deadly ring. By the time Monsignor Mancini realized Scorn was aiming the stun gun at Lovie, he barely had time to leap in front of her.

There was a splash of blood. The next thing Lovie knew, she and Father Paul were catching the monsignor to prevent him from hitting the floor too hard. By the time they lowered him to the floor, doctors Kruel and Minh were there to assist.

"What's my damage?" the monsignor asked, trying to conceal his pain with a smile.

"The stunner grazed you, but the ruptures are irreparable," said Dr. Minh.

"— Father Paul, your advocacy for the children required considerable courage. I should have thanked you long before now," said the monsignor. "Church leaders have a long road to travel before we merit forgiveness from God and the children."

"...then you and I will travel that road together," said Father Paul, trying desperately to give the monsignor hope and the motivation to survive this ordeal.

Lovie was consoling the monsignor as well. She was holding his hand but her heart felt heavy because she'd never apologized for her terrible accusations. She said, "I didn't get a chance to tell you I was sorry—" when *'wham'* came the sound from a powerful kick to her lower back.

"Ow...!" Lovie screamed from the pain. Instinctively, she placed her hands on the injury and when she looked up and saw that Konnerman was her attacker, she knew it was time to fight for her life.

"I'm here to finish what the good doctor started," said Konnerman, backing up and beckoning Lovie to join her in the center of the floor.

"Go get her Slippery Eel", Raquel yelled with her fist balled tight while she threw left and right jabs at a make-believe opponent.

42. The Faceoff

It didn't take much urging to rile Lovie. Father Paul took hold of the monsignor so Lovie could accept the challenge. They met in the center of the studio and began moving in a circle while they sized each other up. Each flinched to get the other to react so they could judge each other's reaction time. That's when Lovie slapped Konnerman hard to the face.

Lovie struck Konnerman as soon as she realized her reflexes were faster. It was an open-handed strike, hard to the jaw. The blow humiliated Konnerman but before she could react, Lovie slapped her again.

Konnerman became infuriated like a raging bull complete with smoke coming out of her nostrils. Her nose was opened wide enough for a truck to do a U-Turn inside. She extended her arms and closed in on Lovie, backing her into a corner. Lovie knew what was happening but whenever she tried to escape, Konnerman threw her back into the corner.

Konnerman was counting on the superior hardness of her head to hurt Lovie when she delivered a shattering head butt to Lovie's forehead. The impact staggered Lovie and the sound of it traveled across the room making everyone cringed.

Immediately, Konnerman grabbed Lovie with a bear-hug which made Lovie's arms useless. Then she delivered a series of head butts, "Let's see you slip out of my grip Ms. Eel," she said loud enough for everyone in the studio to hear.

Although Lovie was unable to use her arms there was nothing wrong with the teeth she used to bite off enough of Konnerman's ear to break loose.

Konnerman screamed, grabbed what was left of her ear and stumbled backwards. Now Lovie was free but she was too dizzy to protect herself. There she stood, vulnerable like a deer blinded by the headlights of an oncoming car. But in this case Konnerman was the car, and she was running straight for Lovie. When she got close enough, she leaped through the air and landed horizontally across Lovie's torso. The impact forced them to crash on top of tables and chairs that broke their falls and prevented Konnerman from landing on top of Lovie.

Konnerman picked up a metal folding chair and walloped Lovie several times across the head and shoulders with it. Lovie collapsed to the floor in a sitting position. Konnerman circled behind her and placed her left arm around Lovie's neck and closed it like a vise.

Lovie was now in the grasp of the sleeper hold—a grip that prevents the carotid artery from supplying blood to the brain. Immediately, Lovie began to feel groggy.

Suddenly something good and unexpected happened. Raquel ran out of the kitchen like a Tasmanian terrorist with the corkscrew in her hand, "Let her go or I will stab you, so help me...you daughter of a witch!"

Then she plunged the corkscrew into Konnerman's arm. It got her attention, but not enough for Konnerman to loosen her grip. Instead, she stared at Raquel with pure hatred in her eyes. Raquel started turning the corkscrew into Konnerman's muscles. It must have hurt, but she did not loosen her grip. Lovie's eyes were closing fast.

Raquel turned the corkscrew again, this time she applied the full weight of her body as it plunged deeper into Konnerman's flesh. Blood squirted a gusher. Konnerman yelled an agonized scream and let go of Lovie.

Raquel wisely took flight and Konnerman took off after her.

Lovie was free again, but she needed a few moments to get her blood circulating. While she watched Raquel run for her life, Lovie decided to use on Konnerman the lesson she had taught her own mother. Even though she was still light-headed, Lovie managed to stand and holler at Konnerman, "Are you chasing her? Or are you running away from me, you bitch?"

Konnerman stopped the pursuit and turned to face Lovie. Then the two combatants ran toward each other. Konnerman tried to connect with a powerful bolo punch, but this time Lovie was too fast. She grabbed Konnerman's left arm with both hands and immediately applied force to the pressure points in her wrist. The results were immediate. Konnerman's brain became confused. She passed out instantly and she would have hit the floor had not Lovie kept her upright.

Working fast, Lovie took a standing position behind Konnerman so they stood back to back. Lovie reached over her own shoulder, pulling Konnerman's head tightly over it. Now Konnerman's head is on one side of Lovie's shoulder, her body on the other side. Suddenly, Lovie dropped to one knee with such force that the impact caused Konnerman's neck to separate from her body with a *'snap'* loud enough to make everyone flinch…and the fight was over.

43. Revelations

Lovie took a long look at Konnerman's dead body before she looked in the direction of Dr. Scorn. "One down and one to go," she said. Lovie was tired. She was breathing heavy and close to exhaustion as she walked toward him with bad intentions.

Dr. Scorn sought protection, hiding behind one of the large Newbie bodyguards.

"Don't kill him for Mike," said the monsignor, but Lovie ignored him and kept walking toward Dr. Scorn.

"...and don't kill him for me, either," added the monsignor. Still, his plea did not deter her.

"Konnerman gave you no option," said the monsignor. "It was kill or be killed. But now the choice is yours...so use it for a civilized purpose." But Lovie continued walking toward her prey.

"You are more than a vigilante," said Father Paul. "For Christ sakes, if you're going to do something in Michael's name, at least make it meaningful."

Lovie stopped and looked at the holy men. "I hear you, but my hatred demands action. I've worked too hard and suffered too much to get to him," she stammered, aiming her fist at Scorn.

"I'm sorry the Church let you down when your family came to us for answers," said the monsignor. "We didn't know the answer back then. And some of my brethren gave opinions rather than admit we didn't know. But now it's time for all of us to make amends."

No sweeter words had Lovie heard. The monsignor's remarks were a catharsis that purged her emotional state. Her mind became clearer. Pity replaced hatred. Justice replaced revenge. But most important, deciding not to kill Dr. Scorn freed the puppet from the puppeteer.

At last, Lovie was free to pursue her personal goals. Now she could publish her thesis and help humanity prepare for the Gender Merger.

44. Arrivederci

The anxiety in Father's Paul's voice travelled across the room like a 911 emergency call.

"Lovie, the monsignor needs to speak with you right away." He waved for her to hurry, and she ran to get there.

"Do you believe in angels?" the monsignor asked her.

"I'm mad at you," said Lovie. "Why did you sacrifice yourself for me? You know I don't like you," she said, unable to hold back her tears.

"Liar, liar…" said the monsignor, wiping one of her tears away with his finger.

Lovie grabbed his hand and kissed it.

"I'm cold. I feel a chill coming on…," said the monsignor. And everyone around him knew this was the end of his days.

The monsignor died peacefully, resting in Lovie's arms. She looked at Dr. Scorn with pure disgust. Scorn stayed behind the guard, but his face remained visible. With the weight of the monsignor's dead body heavy in her arms, Lovie pointed both index fingers at Scorn and slowly took aim at him. "Pow…pow…pow!" she said, with highly animated recoils.

Scorn grimaced, knowing that he was destined to hang from the gallows. Then suddenly, as if on cue the ocean erupted from an underwater explosion that destroyed the United Nation's submarine and gave birth to a new age of reason.

And the human race continued...

www.ingramcontent.com/pod-product-compliance
Lightning Source LLC
Chambersburg PA
CBHW030927180526
45163CB00002B/488
* 9 7 8 1 4 7 7 4 7 2 3 8 5 *